Palgrave Studies in the History of Science and Technology

Designed to bridge the gap between the history of science and the history of technology, this series publishes the best new work by promising and accomplished authors in both areas. In particular, it offers historical perspectives on issues of current and ongoing concern, provides international and global perspectives on scientific issues, and encourages productive communication between historians and practicing scientists.

More information about this series at
http://www.palgrave.com/gp/series/14581

Morris Low

Visualizing Nuclear Power in Japan

A Trip to the Reactor

Morris Low
School of Historical and Philosophical Inquiry
University of Queensland
Brisbane, QLD, Australia

Palgrave Studies in the History of Science and Technology
ISBN 978-3-030-47200-9 ISBN 978-3-030-47198-9 (eBook)
https://doi.org/10.1007/978-3-030-47198-9

This Palgrave Macmillan imprint is published by the registered company Springer Nature Switzerland AG.
The registered company address is: Gewerbestrasse 11, 6330 Cham, Switzerland

To my mother, Eileen Low *(1932–2019)*

Preface

This book explains why the Japanese people embraced nuclear power despite having witnessed its destructive force in Hiroshima and Nagasaki. It highlights the importance of the media, exhibitions, films and tours in helping to achieve a relative consensus regarding the need for the development of nuclear power that ultimately facilitated the transfer of American nuclear technology. The idea of atoms for peace served to separate the good atom from the bad.

The book reflects my longstanding interests in the history of Japanese science and technology, visual culture and issues relating to identity. They are intertwined and it is only by unravelling them that we can truly come to understand Japan's post-war trajectory. Through various projects, I have explored how Japanese identity was increasingly tied to the promotion of science, technology and economic growth. This book complements that work.

Without access to libraries and archives, the book could not have been written. I particularly wish to thank the University of Queensland Library, the National Library of Australia, the National Diet Library, Tokyo, and the US National Archives and Records Administration, College Park, Maryland, USA.

Most of the book was written in my office in the School of Historical and Philosophical Inquiry, University of Queensland. The school has provided me with a congenial working environment. I began research during my sabbatical in 2013. In 2015, I received a Queensland Program for Japanese Education travel grant which enabled me to gather source material in Japan. In 2016, I was awarded an Institute for Advanced Studies in the Humanities (IASH) Teaching Relief Fellowship for Senior Scholars which provided me with time to write up my research. Another period of sabbatical in the second half of 2017 allowed me to make further progress on the manuscript.

I wish to acknowledge my mentors, colleagues, friends and family who have made it possible for me to see this book through to completion. They include Professors Roy MacLeod, Tessa Morris-Suzuki, Robert Kargon, Yoshimi Shunya and the late Professors Nakayama Shigeru and Yoshioka Hitoshi. Professor Peter Harrison, Director, IASH, has been important in promoting the history of science at the University of Queensland through seminars and conferences. I have also benefited from attending seminars on visual politics organized by Professor Roland Bleiker, School of Political Science and International Studies.

As readers will discover, I have learnt much from the work of many scholars who are cited in the pages that follow. I would particularly like to thank an anonymous reader who provided constructive comments that enabled me to produce a more balanced account of visual representations of nuclear power.

I must thank Emily Russell, Publisher, History, at Palgrave Macmillan in London for seeking me out and meeting with me in Brisbane. Thanks also to Megan Laddusaw, Editor, History, at Palgrave Macmillan in New York for taking this project on. I would like to express my gratitude to the editors of the Palgrave Studies in the History of Science and Technology for including this book in their fine series and to Vinoth Kuppan and M. Vipinkumar for putting this book into production.

Brisbane, QLD, Australia
February 2020

Morris Low

Contents

About the Author

Morris Low is Associate Professor of Japanese History at the University of Queensland in Australia. He is co-author of *World's Fairs on the Eve of War* (2015), *East Asia Beyond the History Wars* (2013, 2015), *Urban Modernity* (2010) and *Science, Technology and Society in Contemporary Japan* (1999). He is author of *Japan on Display* (2006, 2012) and *Science and the Building of a New Japan* (Palgrave Macmillan, 2005). He edited *Building a Modern Japan: Science, Technology, and Medicine in the Meiji Era and Beyond* (Palgrave Macmillan, 2005) and co-edited *Asian Masculinities*(2003,2011).ORCIDID:orcid.org/0000-0001-7898-9523.

CHAPTER 1

Introduction: Visualizing Nuclear Power in Japan

Background

The Fukushima nuclear disaster that began on 11 March 2011 was a critical moment in Japanese history. It prompted many people throughout the world to ask why the Japanese people had embraced nuclear power and maintained faith in the energy source, despite having experienced its destructive force in Hiroshima and Nagasaki. This book seeks to answer that question by examining how the Japanese people have seen nuclear power. Atomic energy is not easily visible to the eye unless it is used in some way. We can certainly see its destructive force in the form of a mushroom cloud. It can also be seen in the emblem of the International Atomic Energy Agency (IAEA) which was established in 1957 to promote "atoms for peace": an atom encircled by laurels. The IAEA symbol tends to underplay the death and destruction associated with the bomb, portraying the uses of atomic energy in terms of a simplistic binary opposition.

In the wake of the Fukushima nuclear disaster and amid revelations that components of nuclear weapons have secretly been stored at American bases in Japan in the past, we have come to appreciate how the Japanese people have been encouraged to see nuclear power in a positive light to facilitate the introduction of civilian nuclear power in Japan and at the same time to ensure that Japan has some technical know-how should it ever seek to arm itself with nuclear weapons. This book is about that

M. Low, *Visualizing Nuclear Power in Japan*, Palgrave Studies in the History of Science and Technology,
https://doi.org/10.1007/978-3-030-47198-9_1

journey which culminated in the introduction of British and American nuclear reactor technology and Japanese efforts to make nuclear power its own.

This book seeks to make a contribution to the history of science and the emerging fields of sensory history[1] and visual politics.[2] The story of Plato's cave reminds us how contingent our perceptions can be and the danger of equating what we see with reality. It is this illusion of authenticity that makes the visual so powerful and lends itself to appropriation by government discourses that purport to make knowledge claims. This book highlights the active role of the visual in understanding post-war Japanese history. It was not only through hard work and effort that Japan was rebuilt but also through the senses that the Japanese people grappled with the new post-war world. Any study of visuality must pay heed to not only what was seen but also what was hidden.

Outline of Chapters

Chapter 2 examines how visuality was a consideration in the use of the atomic bomb on Japan. The US government helped shape public perceptions of who in Japan was to blame for the war and the use of the atomic bomb on Hiroshima and Nagasaki was meant to send a strong visual message to Japan, providing the emperor and wartime leaders with a moral justification for surrender. A narrative quickly took hold among the Japanese people that linked deficiencies in science and technology to Japan's defeat in 1945. During the US-led Allied Occupation of Japan (1945–1952), critique of the use of the atomic bomb was discouraged. Physicists such as Yukawa Hideki[3] possessed the specialist knowledge that could make the workings of the atom visible but they were the exception. Anyone who has seen photographs of the effects of the atomic bomb in Hiroshima and Nagasaki would be aware of their affective power. The disfigured and dead bodies of the victims are difficult to view without a sense of horror and revulsion. But the ordinary Japanese person would not see such images of human suffering until the Occupation was well and truly over. Not long after the bombs were dropped, how the Japanese saw the atom was heavily controlled. The suppression of photographs and film footage of the human toll at Hiroshima and Nagasaki shows how politically sensitive representations of nuclear power could be. Censorship of public discussion in the mass media[4] occurred alongside Occupation attempts to re-educate the Japanese in a highly visual manner using exhibits to promote various narratives of how the Japanese would live in a new, more democratic and scientific Japan. Despite censorship, the Occupation was marked by a boom in the publication of popular science magazines.[5] As a result, the atomic bomb was seen more abstractly and more in terms of the power of the atom and how it might be harnessed. Occupation efforts and post-war campaigns to promote US-Japan relations and atoms for peace were effective, given the relative absence of images that reminded the public of the cost to human life.

This book differs from previous accounts of the history of nuclear power in Japan in terms of the attention given to exhibitions, events and representations. While some readers may be familiar with aspects of that story, this book introduces readers to two hitherto neglected figures who throw light on the role of the USA. The first is Frances Baker who worked in the Exhibits Branch, Information Division, Civil Information and Education Section, General Headquarters (GHQ), Supreme Commander of the Allied Powers (SCAP). After the Occupation, Baker worked from the US Embassy in Tokyo for the US Information Agency (USIA) or US Information Service (USIS) as it was known in Japan. In 1954, she married and became known as Frances Blakemore. Her activities provide a window to how the mass media, exhibitions, films and major events helped shape public attitudes towards US-Japan relations, "atoms for peace" and more broadly towards the role of science and technology in Japan's future. The second little-known figure is Clark D. Goodman, an American physicist who went to Japan initially on a Fulbright Fellowship, 1954–1955 and then visited again in 1956 and 1957. Goodman's reports provide a valuable window to the development of nuclear power in the 1950s.

The Japanese were actively encouraged after the war to see civilian nuclear power in a positive light and to dismiss concerns about its safety in an earthquake-prone nation. This book reveals how Japanese attitudes were actively shaped by the Japanese, the USA and British governments, as well as by scientists, media, business figures and industry to view the peaceful atom as inextricably linked to Japan's future. All these players gave the peaceful atom meaning. Although official discourses were contested by concerned citizens, artists and scholars, it is nevertheless striking how the Japanese people have, when surveyed in the past, distinguished between civilian and military nuclear power as if the two were unconnected. By examining forms of visual display which were used to construct knowledge about the atom and how reactors worked, we will come to understand how the Japanese public were actively manipulated. There is a direct relationship between what people know and what they are shown. This book has the simple premise that what people were permitted to see after World War II shaped public attitudes towards the use of civilian nuclear power.

This book argues that public attitudes to nuclear power were shaped by strong interactions between representations and discourses. Despite continuing anxieties about nuclear weapons, a narrative formed that Japan would turn its back on nuclear weapons (the bad atom) and embrace civilian nuclear power. The scientific nature of the discourse around nuclear power created an illusion that it was objective and that nuclear power was safe. It was also part of an American-inspired dream where science could lead Japan. One prime example of how exhibitions promoted US-Japan relations was the 1950 America Fair which Baker helped plan. It is discussed in Chap. 2. The America Fair reflects how after the war, the USA strived to help Japan rebuild its economy and to ensure that its future was aligned with theirs.

The fair which was held in Nishinomiya city, near Kobe and Osaka, permitted Japanese to gain a taste of what it would be like to descend from a Pan Am plane in New York City and see the Statue of Liberty, albeit in scaled-down form. The fair was a milestone in the post-war Americanization of Japan, presenting the Japanese with reproductions and imitations of well-known monuments, historic sites and scenes. Those who visited the imitation White House at the Fair and saw a replica of Mount Rushmore knew what they saw was not the real thing but they were willing to momentarily suspend disbelief and entertain the idea that they could enjoy and live the American dream in Japan. It is the contention of this book that how nuclear power was made visible to the Japanese people also involved a suspension of disbelief which shaped how they saw themselves, their future and the atom.

Exhibitions helped project images of the new Japan. The Japan Trade and Industry Fair known as the Kobe Fair was also held in 1950. It highlighted how Japan had made progress in foreign trade and industry. In the Culture Hall, visitors would learn about the Atomic Age and the contributions made by Japan's first Nobel laureate in physics, Yukawa Hideki. And in the following year, Frances Baker was involved in the "Democratization of Japan" exhibition held in San Francisco in September 1951 on the occasion of the signing of the peace treaty.

At the same time as Frances Baker was producing wartime propaganda in Hawaii and then overseeing American-sponsored exhibitions in Japan, we see in Chap. 3, how the Japanese artist Akamatsu Toshiko illustrated patriotic children's books in Japan and then started painting the famous Hiroshima panels with her husband Maruki Iri from the late 1940s. The mural-sized painting decried war and the use of nuclear weapons at Hiroshima and Nagasaki. They toured the nation and would also be exhibited throughout the world.

Chapter 4 explains how as early as 1950, some American politicians sought to link the development of atomic energy and economic aid. In both Japan and the USA, there were calls for helping Japan to exploit atomic energy given what the Japanese had experienced at Hiroshima and Nagasaki. US President Dwight D. Eisenhower's "Atoms for Peace" speech at the UN in December 1953 flagged a new policy for the promotion of peaceful applications of nuclear technology which the Japanese government was eager to take advantage of. "Atoms for Peace" was not just a construct but a reflection of how many Americans, including politicians and physicists, saw atomic energy as an "agent of redemption"[6] that could enable the atomic bomb to become a source of energy and prosperity for the peoples of the world.

In Chap. 5, we examine how a sense of victimhood was exacerbated in the aftermath of the Lucky Dragon Incident in March 1954 when crew members on a tuna fishing boat were exposed to radioactive fallout. The ensuing controversy and panic about contaminated tuna incited nationwide concern about the dangers of radiation and nuclear weapons. It was a turning point, heralding more discussion of the dangers of the bomb. At the same time, though, focus on the Lucky Dragon served to hide the fate of other fishermen on other vessels that had also been affected. In the aftermath of the incident, we can point to strong efforts to promote US-Japan relations and "atoms for peace" through films, trade fairs and exhibitions which served to influence public attitudes by structuring and organizing how the Japanese people saw nuclear power. The *Family of Man* exhibition that toured Japan in 1956 emphasized the universality of humanity and sought to downplay what had occurred at Hiroshima and Nagasaki.

The Japanese public avidly consumed the spectacle of atomic energy at the *Atoms for Peace* exhibits that toured Japan in 1955 and 1956. The exhibitions used models, artefacts, films and information panels to provide immersive settings in which people could learn about nuclear power. Combined with extensive press coverage, there was an air of excitement created about what the peaceful atom could offer Japan. In this way, the

Japanese and US governments, media outlets and business men delineated what the Japanese people saw and didn't see. In the background, though, was knowledge of what occurred at Hiroshima and Nagasaki. The removal of photographs depicting the atomic bombing of Hiroshima and Nagasaki in *The Family of Man* exhibition in 1956 shows how even after the Occupation, images of the effects of the atomic bomb would continue to be controversial. Newly married Frances Blakemore would also be involved in this exhibition but to what extent she was involved in the removal of what was a photo mural is not clear. Initially hidden from the gaze of the Emperor, the eventual removal was prompted not only out of concern from the Emperor's person but also out of fear that it might stir up political issues relating to the Emperor's wartime responsibility and ultimately adversely affect the US-Japan relationship. The exhibition is thus significant in terms of both what was included and excluded. The selection of photographs reflected not only aesthetic choices but political considerations as well.

A close-up photograph by Yamahata Yōsuke of a young boy taken at Nagasaki in 1945 that remained in the exhibition shows how some images are more powerful than others. It was the sole image left to represent what had taken place at Hiroshima and Nagasaki. The boy's blank, emotionless face powerfully conveyed a sense of shock that the removed photographs also sought to communicate. This is testimony to the emotional impact of visual media but also to their political dimensions. The image elicits the sympathy of viewers more so than a more distant photograph of the same child taken by the photographer.

Chapter 6 discusses six nuclear films including Joseph Krumgold's 1954–1955 documentary *Yukawa Story* which describes how the Nobel Prize-winning Japanese theoretical physicist Yukawa Hideki took up the position of visiting professor at Columbia University after the war, after having visited the Institute for Advanced Study at Princeton. It shows how he and his family served to act as a bridge between the USA and Japan and tradition and modernity. There is only tangential reference to the atomic bomb in the film when a gong bowl is struck. In contrast, the original *Gojira* (*Godzilla*) film which was also released in 1954 sought to convey a more explicit message about the dangers of science. It can be argued that rather than the mushroom cloud which reflects the vantage point of the Americans, it is the mutant monster Godzilla who represents the Japanese view of the destructive force of the bomb in more palatable form. The Japanese projected their fears of the nuclear on to Godzilla who from time to time would rear his head in the original film and in many sequels. In Kurosawa Akira's 1955 film *Ikimono no kiroku* (*I Live in Fear: Record of a Living Being*), there is no monster but only the destructive, psychological effects of fear of nuclear weapons that cripples the main character, Nakajima Kiichi, and prompts him to seek to migrate to Brazil.

Chapter 7 examines the dreams regarding atoms for peace expressed in the media and how 1956 was a key year in terms of the establishment of nuclear infrastructure. The USIS-sponsored *Atoms for Peace* exhibition continued to tour major cities, including Hiroshima and smaller exhibits were shown in regional locations. It was in March–April that Clark Goodman visited Japan at the request of the US State Department. Shortly after, Sir Christopher Hinton visited Japan where he promoted the Calder Hall reactor that he had helped construct and which would be completed later that year. We ultimately see tensions between Americans and the British regarding the merits of their respective nuclear technologies, and there would also be US-Soviet rivalry that could be seen at the 1958 Japan International Trade Fair held in Osaka, as well as in Brussels at Expo '58 which was dubbed the world's first Atomic Fair. By that time, polls in Japan would show increasing public support for the peaceful uses of atomic energy.

Thanks to the many exhibitions, more and more Japanese were able to see what a nuclear reactor looked like and what a nuclear-powered future might hold for them. Chapter 8 focuses on the increasing media coverage of the experimental nuclear reactor facilities at Tōkai-mura and visits by tour groups and individuals. These facilities represented a type of industrial sublime—a combination of both fear of the tamed atom and a sense of wonder at what Japan was able to achieve. In the 1960s, school trips to Tōkai-mura enabled young Japanese to have more of a corporeal experience of where Japan was heading. Prior to viewing nuclear facilities, students would often enter visitor centres and the like where they would add context to what they would encounter. This more embodied experience, albeit in a directed way, made for a heightened experience of what nuclear power had to offer.

Science films provided a visual record of progress made on Japan's first Japanese nuclear power plant. The British Calder Hall type reactor was the centrepiece of the power station at Tōkai-mura and reached criticality on 4 May 1965. Images of construction were woven into wider narratives about Japanese collective identity and how Japan's future would be led by advances in science and technology. Tour groups flocked to Tōkai-mura to partake of this dream. The media was active in presenting images and stories of how science, technology and everyday life would be interwoven—films provided moving images and exhibitions presented the future in three-dimensional form. Site visits by school groups to Tōkai-mura helped promote a myth of safety.

The introduction of a British Calder Hall reactor helps us understand how civilian and military use could be conflated. Despite the "atoms for peace" narrative and the adoption of an anti-nuclear weapons stance, Cold War fears meant that Japan secretly acquiesced to US nuclear weapons being installed at Okinawa before its return to Japan in 1972. Politicians also turned a blind eye to the storage of components on US bases in Japan.[7] Conservative politicians sought to ensure that Japan should possess the technical capability to build nuclear weapons and the necessary plutonium should it wish to go down that path in the future.

Chapter 9 provides details of how the 1964 Tokyo Olympics and 1970 Osaka Expo provided opportunities to promote the narrative that Japan, a nation that had been the first to experience the devastating use of nuclear weapons in wartime, had been rebuilt. The message was that it had turned its back on its wartime past and re-joined the world community. This narrative was literally played out and performed at these events. At the Olympics, Sakai Yoshinori was the final torch bearer at the opening ceremony and he lit the cauldron. The choice of Sakai to do this was carefully thought through. He had been born not far from Hiroshima on the fateful day of 6 August 1945. Similarly, as if to signal that the atom was now put to peaceful use and a central element of Japan's future, the 1970 Osaka Expo was powered by newly opened American-designed nuclear reactors.

Things went awry in 2011 when the Fukushima nuclear disaster put paid to the myth of safety. A tsunami hit the Fukushima Daiichi Nuclear Power Plant as a result of the Tohoku earthquake on 11 March of that year. The whole discourse about nuclear power in Japan was disrupted and the nation is still grappling with the dilemma of whether to restart reactors or somehow replace nuclear power with alternative energy sources. To provide a window to this debate, this chapter also examines representations of nuclear power since Fukushima. The film *Shin Godzilla*, also known as *Godzilla Resurgence* (2016), shows how the Japanese now see nuclear power in a different light. The imagined future powered by the mighty atom that now lies in tatters.

Today, it is possible to see the sights of the Fukushima Daiichi nuclear plant via a computer or a smart phone. Actual presence is not necessary to experience what occurs when nuclear power goes wrong. Increasingly what we see is mediated by technologies. The ubiquitous nature of media in the twenty-first century means that at any one time we now have access to a variety of competing discourses and viewpoints. The way in which nuclear power was made visible and knowable in the past is less tenable and open to contestation.

Where does this leave Japanese national identity in the twenty-first century? There had been hopes after World War II that a new Japan could be constructed, building on the nation's prowess in science and technology. This more future-oriented Japan would not dwell on its wartime past. Reconstruction of the economy took priority and by the 1964 Olympics, Japan was well on its way to becoming a major economic power. A national imaginary emerged that showcased its achievements in science and technology. This nexus between science and national identity can be seen in the history of the introduction of nuclear power and the rise of popular culture icons such as Astro Boy and Godzilla which would provide the affective bond between the Japanese state and government efforts to promote Atoms for Peace. These and other representations of nuclear power discussed in this book served to mediate the relationship between the Japanese and US governments, and the Japanese people. Fukushima disrupted narratives about national identity. The nation once again turns to major events such as the Olympic Games and a World Expo to rally the population and show the world that it has reconstructed, recovered and keen to face the future. Science and technology will once again be deployed but arguably more attentive to the needs of the Japanese people. In Chap. 10, we reflect on what is covered in the book and how visual culture offers ways in which to bring about change.

Notes

1. Mark M. Smith, *Sensory History* (Oxford: Berg, 2007); Kerim Yasar, *Electrified Voices: How the Telephone, Phonograph, and Radio Shaped Modern Japan, 1868–1945* (New York: Columbia University Press, 2018).
2. Roland Bleiker, ed., *Visual Global Politics* (Abingdon, Oxon: Routledge, 2018).
3. Japanese names in this book are given in the Japanese order of family name followed by given name. In the case of names of Japanese authors whose

texts have been published in English, they are cited in endnotes according to the order given in the publication.

4. Monica Braw, *The Atomic Bomb Suppressed: American Censorship in Occupied Japan* (Armonk, N.Y.: M.E. Sharpe, 1991).
5. Yukio Wakamatsu, "The Mushrooming of Popular Science Magazines," in *A Social History of Science and Technology in Contemporary Japan*, Vol. 1, The Occupation Period: 1945–1952, eds. Shigeru Nakayama, Kunio Gotō and Hitoshi Yoshioka (Melbourne: Trans Pacific Press, 2001), 516–532.
6. Susan M. Lindee, *Suffering Made Real: American Science and Survivors at Hiroshima* (Chicago: University of Chicago Press, 1994), 15.
7. William Burr, Barbara Elias and Robert Wampler, eds., *Nuclear Weapons on Okinawa Declassified December 2015, Photos Available Since 1990*, Briefing Book No. 541 (Washington, D.C.: National Security Archive, George Washington University, 2016), https://nsarchive.gwu.edu/briefing-book/japan-nuclear-vault/2016-02-19/nuclear-weapons-okinawa-declassified-december-2015-photos-available-1990

Bibliography

Bleiker, Roland, ed. *Visual Global Politics*. Abingdon, Oxon: Routledge, 2018.

Braw, Monica. *The Atomic Bomb Suppressed: American Censorship in Occupied Japan*. Armonk, N.Y.: M.E. Sharpe, 1991.

Burr, William, Barbara Elias and Robert Wampler, eds. *Nuclear Weapons on Okinawa Declassified December 2015, Photos Available Since 1990*, Briefing Book No. 541. Washington, D.C.: National Security Archive, George Washington University, 2016. https://nsarchive.gwu.edu/briefing-book/japan-nuclear-vault/2016-02-19/nuclear-weapons-okinawa-declassified-december-2015-photos-available-1990.

Lindee, Susan M. *Suffering Made Real: American Science and Survivors at Hiroshima*. Chicago: University of Chicago Press, 1994.

Smith, Mark M. *Sensory History*. Oxford: Berg, 2007.

Wakamatsu, Yukio. "The Mushrooming of Popular Science Magazines." In *A Social History of Science and Technology in Contemporary Japan*, Vol. 1, The Occupation Period: 1945–1952, edited by Shigeru Nakayama, Kunio Gotō and Hitoshi Yoshioka), 516–532. Melbourne: Trans Pacific Press, 2001.

Yasar, Kerim. *Electrified Voices: How the Telephone, Phonograph, and Radio Shaped Modern Japan, 1868–1945*. New York: Columbia University Press, 2018.

CHAPTER 2

Before and After Hiroshima

Before Hiroshima

During World War II, many people throughout the world associated Japan's war effort and atrocities with its emperor. *Fortune* magazine, however, saw things differently and sought to convince the public in a special issue on Japan that was published in April 1944 that combined text with effective images. The article "Who Runs the Emperor?" argued that it was the militarists running Japan who were responsible and that the emperor was merely a symbol. To visually reinforce this argument and sway public opinion, a strip of photographs ran along the bottom of the first page of the article with the heading "Militarists like These Rule Japan." The mugshot like photographs showed the heads of Prime Minister Tōjō Hideki, Field Marshal Terauchi Hisaichi, Commander-in-Chief of the Combined Fleet Koga Mineichi and the Navy Minister Shimada Shigetarō. The caption explained that

> The Son of Heaven is a little man weighed down by the trappings of empire. His personal power: none.... His function: to be used by power groups headed by men like the four hard-faced militarists below.[1]

It ran alongside a full-page photo showing the small figure of the emperor in all his finery, sitting in a fancy carriage which took up more than half of the page. The article argued that "His value to Japan is his value as a myth,

M. Low, *Visualizing Nuclear Power in Japan*, Palgrave Studies in the History of Science and Technology,
https://doi.org/10.1007/978-3-030-47198-9_2

and the myth has been artfully assembled and cleverly merchandised."[2] The layout was repeated in the following two pages, only this time the headings of the four photographs running along the bottom were "Civilians Still Have Influence" and "Horn-Rimmed Empire Builders," respectively. The caption of the latter somehow associated wearing spectacles with having "spurred Japanese expansion," reinforcing the negative stereotype of Japanese as bespectacled, bucktoothed Asians.[3]

Another article pondered "What to do with Japan?"[4] It was followed by a full-page advertisement for the new low-pressure reinforced plastic products made of Valinite which was being promoted by the furniture manufacturing company Virginia-Lincoln Corp. in Marion, Virginia. The ad appropriated Lewis Carroll's character of Alice from *Alice's Adventures in Wonderland* (1865) and showed her peering "through the looking glass" at Valinite products into the future. It exclaimed:

> Alice, your wonderland has grown up! On V-Day, war proven Valinite will return from the battle fronts to serve America in the building of a better tomorrow…. Our Postwar Planning Bureau will be glad to help you with your problems.[5]

By appropriating the story of *Alice in Wonderland* which was particularly popular with children, the advertisement effectively used children's culture for propaganda purposes. It explained how Valinite was being used for the war effort. It showed how plastics were patriotic and would help make people's lives better after the war. *Fortune* and indeed American private enterprise already seemed to be preparing for the years that were to follow. Satisfying consumer desires for products that were inexpensive, lightweight and disposable[6] would drive economic growth in peacetime both in the USA and in Japan. And the *Alice in Wonderland* narrative would later be used to sell other things such as nuclear power.

The editors of the special issue of *Fortune* acknowledged "the many private individuals and official agencies" who made the special issue possible including "several sources who may not be named." The US State Department is likely to have been among them.[7] There is strong correlation between the advice given by the historian Hugh Borton who was working for the State Department and advising on post-surrender policy including the future role of the emperor. In October 1943, the State Department established the Interdivisional Area Committee on the Far East which oversaw approval of policy documents for endorsement as

State Department policy. Borton and his colleagues helped prepare documents on Japan for their consideration and these were ultimately made available to the Postwar Programs Committee which was established in January 1944 as the highest policymaking body within the department.[8] Borton and his colleagues also provided advice to the War Department. He had long argued that after the war, the Japanese people would blame their military leaders and not the emperor for their defeat. What's more, he foresaw that they would continue to see him as a symbol of the nation and that after surrender they would follow whatever instructions he gave them.[9]

Meanwhile, the challenge remained to induce Japanese military forces and Japanese people to surrender. A US Office of Strategic Services memo dated 8 July 1944 referred to a joint Anglo-American plan for psychological warfare against Japan that had just become official policy in May.[10] The plan would capitalize on the perception of the superiority of British and American weapons and United Nations manpower. This, combined with the lack of supplies and low morale among Japanese soldiers, was seen as likely to increase their sense of inferiority. Carefully timed propaganda could have the desired effect of encouraging them to surrender.[11] Among the American propagandists tasked with producing leaflets was Frances Baker (later to be known as Frances Blakemore). Baker was chief illustrator in the art department at the Office of War Information, Central Pacific Operations, Honolulu. Although born and raised in eastern Washington, she went to Japan in 1935, staying there for five years. She escaped to Honolulu just prior to the outbreak of the Pacific War. Among the many leaflets that she produced, there was one on how the Japanese military had underhandedly seized control of the government and were to blame for the war. In this way, the narrative was reinforced to the Japanese people with some 400,000 copies of this one leaflet alone dropped on Japanese cities.[12]

On 31 May 1945, Secretary of War, Henry L. Stimson, chaired a meeting of an eight-member Interim Committee which met with the approval of President Harry S. Truman to provide advice on a range of matters including wartime controls, post-war policy and nuclear weapons. Members included the engineer Vannevar Bush who headed the Office of Scientific Research and Development (OSRD). Bush had been instrumental in persuading the US government on embarking on a program to build an atomic bomb known as the Manhattan Project. He was joined by his OSRD colleague, the physicist and President of MIT, Karl T. Compton, as well as the chemist and President of Harvard University James B. Conant. Four distinguished physicists who had contributed to the Manhattan Project were specially invited to attend the meeting as a scientific advisory panel: J. Robert Oppenheimer, the so-called father of the atomic bomb who headed the Los Alamos Laboratory where the weapon was developed; Enrico Fermi who is credited with building the world's first nuclear reactor and considered as one of the architects of the bomb; Arthur H. Compton (Karl's brother) who headed the Metallurgical Laboratory that produced nuclear reactors that converted uranium into plutonium for use in nuclear weapons; and Ernest O. Lawrence who invented the cyclotron and worked on uranium isotope separation. Four other people attended, including Major General Leslie R. Groves who was director of the Manhattan Project.

Oppenheimer suggested that

> it might be wise for the United States to offer to the world free interchange of information with particular emphasis on the development of peace-time uses. The basic goal of all endeavors in the field should be the enlargement of human welfare. If we were to offer to exchange information before the bomb was actually used, our moral position would be greatly strengthened.[13]

Bush acknowledged that "our tremendous advantage stemmed in large measure from our system of team work and free interchange of information,"[14] but he was concerned about whether the USA could remain permanently ahead if it made research results freely available to the Russians and received nothing in return. In Karl Compton's opinion, "secrets of this nature could not be successfully kept for any period of time and that we could safely share our knowledge and still remain ahead."[15] There was thus recognition even before the bomb was used that the secret of the atomic bomb would become known to others, that the benefits of atomic energy should be made known to others and that the USA would probably stay ahead.

The meeting went on to discuss the effect of the dropping of the atomic bomb on Japan and the will of the Japanese to continue fighting. Although the effect of the dropping of a bomb on an arsenal would not be so different from that which could be achieved by an Air Corps strike using more conventional weapons, Oppenheimer stated that "the visual effect of an atomic bombing would be tremendous. It would be accompanied by a brilliant luminescence which could rise to a height of 10,000 to 20,000 feet."[16] After much discussion, Stimson came to the generally agreed conclusion that the Japanese would not be given any warning and while they should not concentrate on a civilian area, "we should seek to make a profound psychological impression on as many of the inhabitants as possible."[17] The Committee thus recommended military use of the atomic bomb against Japan. Information on atomic energy would not (at least initially) be shared with the rest of the world and the Japanese were not forewarned that the bomb would be used against them.

In a last ditch attempt to prevent this from happening, Arthur H. Compton submitted a memorandum to Stimson on 12 June 1945, on behalf of some key members of the scientific staff of the Metallurgical Laboratory at the University of Chicago, for the attention of the Interim Committee. The document, later known as the Franck Report after James Franck who chaired the group which instigated it, casts doubt on whether the first available, small bombs would be sufficient to break the will of the Japanese to continue fighting, especially given that major cities like Tokyo, Nagoya, Osaka and Kobe had already been largely destroyed by ordinary aerial bombing.[18] It urged the USA to make a technical rather than a military demonstration of the atomic bomb on a deserted island somewhere "before the eyes of representatives of all United Nations"[19] in the hope that it would lead to international agreement that such weapons would be outlawed. Sadly, this plea was to no avail. On 16 June 1945, Oppenheimer, writing on behalf of the four scientist advisory panel that had been invited to the Interim Committee meeting, suggested that "we can propose no technical demonstration likely to bring an end to the war; we see no acceptable alternative to direct military use."[20]

That month's issue of *Fortune* also seemed to agree. It suggested that "What Japan needs is a symbolic act in tune with high and ancient Japanese tragedy" and "what Japan is looking for now is a formula for a 'moral' surrender." The editorial ruled out the possibility of a negotiated peace and any chance that the emperor would provide such a formula himself as "he is only a Name signed to orders issued by others,"[21] despite being the central focus of Japanese loyalty and duty. *Fortune* magazine knew more than it was letting on. And as the historian Borton had predicted in his draft papers advising on the treatment of the emperor, the Japanese people would comply with whatever the emperor would ask of them.[22] Thus, by June 1945, a clear narrative had emerged. A decision had been made for the military use of the atomic bomb, and there was a belief that it would lead to Japan's surrender.

Bombing of Hiroshima

On 6 August 1945, Groves wrote a memorandum to the Chief of Staff, reporting that an atomic bomb had been dropped on Hiroshima where "there was no indication of any POW camp" lest lives of Allied soldiers be at risk. The visible effects of the bomb which had been dropped that morning (Japan time) were judged to have been greater than that of the New Mexico test on 16 July although the flash was deemed not as blinding as the test explosion due to bright sunlight on the day. The mushroom cloud was described evocatively as follows:

> First there was a ball of fire changing in a few seconds to purple clouds and flames boiling and swirling upward. Flash observed just after airplane rolled out of turn. All agreed light was intensely bright and white cloud rose faster than New Mexico test reaching thirty thousand feet in three minutes it was one-third greater diameter. It mushroomed at the top, broke away from column and the column mushroomed again. Cloud was most turbulent. It went at least to forty thousand feet. Flattening across its top at this level.[23]

A high speed camera had captured an excellent record of the bombing of Hiroshima and observer planes had also taken post-strike photographs. But Groves' deputy Major General Thomas F. Farrell would travel to Hiroshima to see for himself. In a memorandum dated 10 September 1945, he reported having undertaken a preliminary inspection, flying over the city and inspecting it in detail on the ground:

> The scene is one of utter devastation, the airplane photographs do not give an adequate presentation of the complete levelling of the City which is substantially destroyed. One there had awful proof of the terrible power of the bomb.[24]

He travelled on to Nagasaki where he reported that the effects of the bomb that had been dropped there in the morning of 9 August (Japan time) are "more spectacular and startling than Hiroshima…The larger distance of destruction of worker houses indicates a blast energy about twice that at Hiroshima."[25]

Farrell's comments suggest that the aerial view of the effects of the atomic bombs dropped on Hiroshima and Nagasaki was insufficient in terms of making visible the "terrible power" of the new weapon. We can turn to the published diary of the physician Hachiya Michihiko for a sense of how it affected the Japanese people below on the ground who fell victim to the bomb. While Hiroshima was far removed from the emperor in Tokyo, those responsible for the safeguarding of the imperial portrait at schools and other government institutions went to great lengths to protect it from harm's way amidst the devastation of the just-bombed city as if the portrait was a proxy for the emperor himself. The group protecting the portrait made its way to the river's edge through the thronging masses shouting "The Emperor's Picture! The Emperor's Picture!" People stood and saluted or bowed and those who couldn't simply offered a prayer on the spot. The crowd showed great deference and respect to the imperial portrait, even as flames enveloped the landscape. They made way for the picture and it was carried safely to the water.[26]

Surrender

The bombing of Hiroshima and Nagasaki provided the visual spectacle that Oppenheimer felt was needed and the psychological impact that Stimson and the Interim Committee sought to deliver. It also provided a symbolic act and a formula for a moral surrender that *Fortune* magazine felt was required. At 12 noon, 15 August 1945 (Tokyo time), a prerecorded speech by Emperor Hirohito was broadcast on radio. In his speech, given in formal classical Japanese, he stated that the government had been instructed to accept the terms of the Potsdam declaration which had demanded the unconditional surrender of the Japanese. An English translation was broadcast overseas at the same time and published in *The New York Times*. It included the following sentences:

> Moreover, the enemy has begun to employ a new and most cruel bomb, the power of which to do damage is, indeed, incalculable, taking the toll of many innocent lives. Should we continue to fight, it would not only result in an ultimate collapse and obliteration of the Japanese nation, but also it would lead to the total extinction of human civilization.[27]

But while the speech was received solemnly by the Japanese people who had difficulty understanding what he was saying but nevertheless understood the gravity of the occasion as they were hearing his voice for the first time. Some Japanese gathered outside the Imperial Palace in Tokyo, knelt on the gravel with head bowed to the ground, feeling ashamed that their efforts had not been sufficient to achieve victory but others felt angry and deceived by its wartime leaders, and for many there was a sense of loss of purpose. The emperor's own son, Crown Prince Akihito, recorded in his diary that Japan had lost the war primarily due to two reasons: Japan had been behind the USA in science and technology, and Japan had lagged behind Americans in terms of working as a group due to individual selfishness.[28]

News of the emperor's speech was received differently in the Allied nations where it was greeted with relief and jubilation. We can get some sense of it by merely looking at the page on which the speech appears in *The New York Times*. Three photographs recorded the situation before and after President Truman's official announcement of Japan's surrender. Anticipating what was to come, a long conga line of civilians and servicemen snaked along on the grass of Lafayette Square in front of the White House. Below that, a photograph showed a jam-packed Market Street in San Francisco after the announcement and another showed sailors in Boston hoisting up a baby above their heads in celebration.

On 15 December 1945, the General Headquarters of the Supreme Commander for the Allied Powers which occupied Japan after the war issued a memorandum for the Japanese government known as the Shinto Directive which ordered the abolition of government support of State Shinto, separation of church and state, and guaranteed freedom of religion in Japan in order to "assist the Japanese people in a rededication of their national life to building a new Japan based upon ideals of perpetual peace and democracy."[29] *The New York Times* reported on Supreme Commander General Douglas MacArthur's order but speculated

> how the ultra-nationalistic and militaristic elements of Shinto can be eliminated as long as an Emperor, whose title to his throne rests upon his alleged descent from divine warrior ancestors, remains as both temporal and spiritual head of the nation.[30]

Thus, a further result of the Shinto Directive was the emperor's declaration of "humanity" on 1 January 1946 whereby he publicly stated that he was not "divine."[31] This arguably created an ideological vacuum which science and technology helped fill. Whereas previously, national identity was strongly tied up with the mythology of the divine origins of the imperial family and supposedly unbroken line of emperors, they could now look elsewhere.[32] Emperor Hirohito had connected the "new and most cruel bomb" to Japan's defeat in his surrender speech, and the lesson that many took from this was that the Japanese had lost due to superior American science and technology rather than any other failings on the part of the Japanese people.[33] This is despite the atomic bomb having been the product of a large team of people from many diverse backgrounds who had worked on the Manhattan Project and who had benefited from the free exchange of knowledge before the war, as the Interim Committee had discussed. With the declaration of humanity, a more human emperor was revealed to the Japanese people and the media helped promote him not only as the people's emperor but also as a scientist himself![34]

After Hiroshima

In the aftermath of the dropping of the atomic bombs, the US Strategic Bombing Survey (USSBS) which had been established in November 1944 by Stimson to examine the effects of strategic air power being used in Europe and extended by President Truman to include the aerial war against Japan also included an examination of the impact of the bombs on Hiroshima and Nagasaki. In addition, there was discussion within the US government regarding the future of nuclear weapons. At a cabinet meeting, Stimson, echoing opinions aired at the Interim Committee meeting referred to earlier, suggested that information about the peaceful uses of atomic energy be shared with the Soviets given that it could prove critical to US-Soviet relations and given that the basic information needed to build a bomb would get out soon enough anyway. Such ideas preceded what would become known as the Atoms for Peace program by several years.[35]

A report from the USSBS Pacific Survey was issued by the chairman's office on 1 July 1946 entitled *Japan's Struggle to End the War*. A typescript version from the Truman Library and Museum states that

> The Hiroshima and Nagasaki atomic bombs did not defeat Japan, nor ... did they persuade Japan to accept unconditional surrender. The Emperor, the lord privy seal, the prime minister, the foreign minister and the navy minister had decided as early as May of 1945 that the war should be ended even if it meant acceptance of defeat on allied terms.[36]

This stands in contrast to the emperor's speech. It can be argued that the use of the atomic bombs provided an opportunity for the emperor and his advisors to craft a narrative with very different lessons for Japan. This narrative would in turn shape how the Japanese people experienced and understood the world, helping them to view their wartime past more as victims rather than aggressor. As for their future, the nation would harness science and technology to deliver material benefits, economic growth and prosperity for the Japanese people. In a resource-poor nation, they would pin their hopes on nuclear power and aligning the nation with the USA would be part of the deal.[37]

Clark Goodman and Harnessing Atomic Energy

During the war, Clark D. Goodman worked for the Office of Scientific Research and Development (OSRD) and was later employed as a senior physicist at Oak Ridge National Laboratory where he worked on nuclear propulsion for naval vessels. He returned to MIT as a member of faculty in 1947 where he had completed a PhD in physics in 1940.[38] Goodman was the editor of *Science and Engineering of Nuclear Power* (1947–1949),[39] a two-volume book which was the first unclassified text on the subject. The second volume was dedicated to the UN Atomic Energy Commission "whose responsibilities include disseminating among nations scientific information for peaceful purposes."[40] Goodman begins his chapter in that volume with a quote from David Lilienthal from September 1948. Lilienthal had contributed to the 1946 *Report on the International Control of Atomic Energy* known as the Acheson-Lilienthal Report. He was also inaugural Chairman of the US Atomic Energy Commission, 1946–1950. In Lilienthal's words

> Nuclear energy is an elemental process which is at once and in one package a key to knowledge, a means of therapy, a source of power and a weapon of devastating destructiveness. And the point is that it is all these things *at once*.[41]

Despite later developments to separate "atoms for peace" from "atoms for war," Lilienthal was stating the obvious. Indeed, Goodman would on that same page state that "the major applications of nuclear energy have been and in all probability will continue for some time to be military in nature."[42] He added that

> Any type of nuclear reactor built for the generation of power is susceptible with only relatively moderate effort, to diversion into the production of

> Pu^{239} and U^{233}. In concentrated form these nuclear fuels are suitable for atomic armament. Thus, danger to world security is inherent in the development of nuclear energy for industrial purposes.[43]

We will meet Goodman again later in this book when he took three years of partial leave from MIT to serve as Assistant Director of the Division of Reactor Development of the US Atomic Energy Commission. It was at this time that he and his wife spent a year in Japan in 1954–1955 on individual Fulbright Fellowships. A subsequent trip to Japan in early 1956 provides a window to the rapid developments occurring in the development of nuclear power that year. Meanwhile, it fell to the CIE, GHQ, SCAP and later the USIS to convince the Japanese that they could safely pursue the peaceful uses of atomic energy against the background of the US-Japan relationship.

Frances Baker and the CIE

During the war, while Goodman was working for the OSRD, Frances Baker participated in psychological warfare support activities against the Japanese through the production of propaganda for the Office of War Information in Honolulu. She had lived in Tokyo from 1935 to 1940, working as an artist and English teacher, and this was a way of utilizing her knowledge of the country and its people and making the most of her skills. After the war, she was keen to return to Japan and took up a position in Tokyo in 1946 as a civilian member of the Civil Information and Education Section (CIE), a staff section of General Headquarters, Supreme Commander for the Allied Powers. CIE had been established in September 1945 with the responsibility for accomplishing Occupation objectives in fields and matters relating to information, education, religion, culture, the fine arts, public opinion and sociological research areas. It was acknowledged that the CIE dealt "largely with intangibles, with matters closely related to the thought and culture patterns of the Japanese people."[44] Much of the Americanization of Japan after the war has been attributed to CIE cultural policies.[45]

The Information Division of CIE carried out information campaigns and it was here, from early 1947 that Baker worked as exhibits officer in the exhibits unit, alongside a dozen Japanese staff members in the Radio Tokyo Building near Hibiya Park.[46] In that key role, Baker continued to help shape Japanese attitudes through the design and production of printed visual material and exhibits to promote the Occupation's goals and campaigns to change Japan. These materials were sent throughout Japan to convey to the Japanese people in schools, factories, department stores and libraries information about the Occupation's democratic reforms including politics, the emancipation of women and civil liberties, labour-management relations and industrial democratization, land reform, the civil code, local autonomy, the role of education boards, and the role of agricultural and fishing cooperatives.[47] The visual materials that were distributed included photographs, cartoons, charts and projector cards. In 1947 alone, CIE generated 460 sets of exhibits. In 1949, some 722 exhibits were created.[48] These figures show the continuities between wartime propaganda and post-war campaigns to influence what the Japanese saw and thought, helping to manage perceptions of the USA through careful selection of information that would be conveyed in exhibitions that would portray the USA and its reforms in a favourable light. This would help promote a desire for things American nationwide.[49]

To get a sense of the wide breadth of her work, we can view Baker's fortnightly report for the period ending 15 November 1949, written under her then name of Frances Baker (she would marry Tokyo-based, American-born lawyer Thomas Blakemore in 1954). By this time, it was known as the Exhibits Branch. Baker reported that the Hibiya Public Library had agreed to display the exhibit on the New York Public Library, and the Radio Tokyo Building would allow the exhibit on Student Legislature to be shown at its entrance. The Sanseidō Book Store in Kanda had agreed to display books that had been translated under the CIE Translation Program.

The Exhibits branch also produced sets of *kamishibai* (paper theatre or picture-story show) which involved a live presenter who would narrate a story in the streets, using hand drawn or printed visual panels to engage the audience. Baker reported that 920 sets of the *kamishibai* "Starting a New Life" had been distributed via prefectural visual aid centres. The branch cooperated with the Japan *Kamishibai* Association and the Tokyo Metropolitan Education Bureau to display one set at the Mitsukoshi Department Store in Shinjuku from 12 to 20 November and two performances were given daily.[50]

Baker estimated that the daily showing of exhibits generated by the office between 1 and 15 November totaled 1983.[51] The many exhibits included one on "Science and Innovation" which was lent to the Yamagata Civil Affairs Team. In addition, 167 photographs were released to the

Photo Officer, Press and Publications Branch, CIE for use by Japanese organizations including newspapers and magazines. In this way, the Exhibits branch helped the Japanese public envisage a new, more democratic Japan that drew on American models.

The Educational Films Unit (EFU) was within the Motion Pictures and Theatrical Branch of the Information Division, CIE. EFU was also important in this endeavour. It sent imported documentary films to Japanese film companies for adaption for Japanese audiences. Yuka Tsuchiya has identified fourteen film titles on science and technology. These included a seventeen-minute film on "Atomic Power" which portrayed the relationship between science and humans in a highly positive light.[52]

The America Fair

In the final years of her time with CIE, Baker participated in two key exhibitions which marked the ending of the Occupation: the America Fair (1950) and the *Democratization of Japan* exhibition (1951). The major organizer of the America Fair was the newspaper *Asahi Shimbun* but Baker contributed as a display advisor on behalf of the CIE. Col. Wiley O'Mohundro, Commanding Officer of the Kinki Civil Affairs Region, and Lt Col Donald R. Nugent, CIE Chief, were among specially invited guests who attended the formal opening ceremony on 17 March. A speech written by Maj. Gen. W.B. Kean, Commanding General of the Twenty-Fifth Infantry Division, Osaka, was read out to guests on his behalf by Captain George Sammet Jr at the ceremony.[53] Lt Col Nugent himself gave a speech on the stage which was the focus of the opening ceremony proceedings, flanked by the US and Japanese flags on the curtain behind. Frances Baker was the only woman on the stage. She sat in prime position near the Japanese flag in an acknowledgement of her work in the months leading up to the opening. When the Fair opened the day after at Nishinomiya near Osaka on 18 March 1950, Mrs Kean cut the opening ribbon at the gate,[54] immediately beyond which visitors could see a model of the White House. The participation of Occupation personnel in these various ways at the America Fair flagged how it was done with their support and approval.[55]

Some 2 million visitors flocked to such exhibits over three months from 18 March through to 11 June. Held at the Nishinomiya Stadium and adjacent areas, the Fair was effectively divided into two sites. The Fair primarily showcased American history and politics, its economy and education system, culture and way of life, with many of the displays showed the benefits of science and technology. For example, on the fourth floor of the main exhibition hall at the stadium, visitors could see how atomic energy could be used for peaceful purposes through the use of medical isotopes and Geiger-Müller counters as diagnostic tools.[56]

A notable feature was the use of panoramas to give the illusion of travelling through time and place to arrive at present-day America. The first site included an indoor panorama showing scenes from US history, and visitors could see how American science had made everyday life easier. There was a display of appliances, a model home and a television pavilion where they could view live television for the first time. Japanese companies also mounted displays in the New Japan Industry Hall to flag the importance of rebuilding the Japanese economy under the watchful gaze of the USA.[57]

The second site provided an outdoor panorama so that visitors could simulate the experience of travelling to the USA. They could board a model of a Pan-American Airlines Stratocruiser airplane as if they were on an actual flight, travelling from San Francisco and visiting iconic destinations such as New York City, Niagara Falls which was complete with a scale model of the Peace Bridge, the Grand Canyon and Yosemite National Park, and even the Capitol Building, home of the US Congress in Washington, DC. They could also boast of having visited a model of the White House, viewed Indian Villages, seen a Hawaiian dance and played at the Little Coney Island. To remember their trip to "America," visitors could have a photograph taken in one minute with the recently released Polaroid Land Camera.[58] The photograph would provide a record, albeit a miniaturized one, of the Fair which in turn was but a model and simulation of America, its past and present.

CIE appears to have had considerable input in helping to plan the Panoramic View of America. It gave visitors an experience that was not unlike visiting Disneyland, complete with a "Little Coney Island" for children with rides. The second site also included a CIE Information Centre for adults and a cinema where all visitors could view American tourist movies to see the real thing. In the official record of the America Fair, Frances Baker is photographed with CIE colleagues examining a model of the panorama. It was a crucial feature of the second site and of the fair itself that was highly successful.[59]

The view of America that was conveyed was bright, light and inspiring. A grade 6 primary school student Nishino Shinji visited the model home at the Fair and was struck by how bright everything was with big windows that let in light and white painted walls and how convenient things were, especially in the kitchen. He was particularly attracted to the domestic electric appliances that were part of everyday life in the USA, from the electric fridge to the coffee maker. This left the impression that people such as his mother would have more time left in their day for recreation purposes such as reading books and listening to the radio. He came to the conclusion that in order for the Japanese to enjoy such a way of life, not only was it a case of having a sufficient income to pay for it but also that the Japanese people themselves needed to study science and make various machines and tools themselves. They needed to become a "*kagakutekina kokumin*" ("scientific or science-oriented people") in order to be able to freely use such things. A third year, junior high school student Ikeda Hisao saw American home life as being democratic and respectful of the individual. He saw it as being both hygienic in its cleanliness and scientific, a wonder to behold. For both young Ikeda and Nishino, science and technology were reflected in the American-style way of life.[60]

By putting it on display, the *Asahi Shimbun* newspaper, with the assistance of government ministries, the National Railway Corporation and the Nishinomiya city office, as well as support from the USIS, helped foster a general preference for American models to emulate, whether it be the way of life, household conveniences or science and technology more broadly. And for those who weren't able to visit the fair in person, the Shōchiku Motion Picture Company, under CIE's supervision, filmed the Fair with the aim of screening it in both Japan and the USA.[61]

The *Nippon Times* on 23 March suggested that the Fair was for "the multitudes in Japan who desire to see and know about America…the next best thing to taking a trip to the land of their longing." The Fair provided "samples of American civilization on display." There were miniature models: an American apartment house with a kitchen outfitted with electric appliances, a farm with a one-man tractor and a middle-class American home to "satisfy the curiosity of those who have but a hazy idea of [American life]."[62] The article made plain that the Japanese would not see the real thing but what was on display would be sufficient to state their "longing." Lt Col Nugent, in his opening speech which was published in the *Nippon Times* in the same issue, thanked the *Asahi Shimbun* for "its endeavor to make more real to the Japanese people the nation which circumstances have made so vitally important to their present and future welfare."[63] He felt that

> this America Fair cannot but have a catalytic influence. It will enlighten, stimulate and even inspire. Some of the things seen here will be remembered and talked about for years to come. Out of observation, analysis and the making of comparisons will develop ideas which may have impacts on all aspects of Japanese life.[64]

In some quite perceptive comments regarding the visualization of America at the Fair, he regretted "the difficulty of representing and making visible those things about a nation and its people which lie behind and transcend the material and therefore visible things."[65] He added that

> Impossible to show and make known so clearly are the spiritual values of the people who make and use these things, their attitudes toward one another, and what moves them to action. These, however, are America just as much as the machines and gadgets and high living standards.[66]

He went on to say that in Communist, totalitarian nations

> there is only disregard for but actual obliteration of the spiritual values which underlie the real progress of mankind. There, the individual exists only to serve the state and to contribute only to its material development.[67]

As much as the Japanese public seemed to embrace American culture and way of life, they were not keen being part of the emerging Cold War. The same issue of *Nippon Times* reported that a United Press survey of Tokyo bureau chiefs of nine large newspapers throughout Japan indicated that the Japanese were not in favour of having foreign bases in Japan after a peace treaty was signed. Some felt that it would violate Japan's constitution and some thought that retaining US bases might invite trouble between Russia and Japan.[68]

The American magazine *Popular Mechanics* made fun of the America Fair in an illustrated article entitled "'Little America' in Japan" in the June 1950 issue. Spread over two pages, the magazine showed an American GI in Japan looking up at a copy of the famous statue of a seated Abraham Lincoln, as well as a kimono-clad woman holding balloons in front of the White House. In addition to the models of famous American buildings, monuments and natural phenomena such as the Grand Canyon, the article noted that there were even exhibits of bingo games and slot machines, such was the extent of the interest in things American. But what seems to have amused the author most was that the famous sculptures on Mouth Rushmore "have been skilfully reproduced, with a strange Oriental look."[69] Indeed, the faces of US presidents George Washington, Thomas Jefferson, Theodore Roosevelt and Abraham Lincoln all seem to have been rendered with decidedly Asian faces. Something does seem to have been added in the process of remaking the famous landmark in Japan. There was slippage in transferring America to Nishinomiya. What we see there is an imagined America.

Hasebe Tadasu, president of the *Asahi Shimbun* Publishing Company and president of the America Fair, addressed the mimetic nature of the fair and acknowledged that many aspects of America on view "are peculiar only to America and cannot be copied by the Japanese." He felt, nevertheless, that it was important for the Japanese people to acquaint themselves with every aspect of America and "to make good use of it for our own future, and at the same time to reflect upon ourselves and our past." To this end, a display of Japanese products had also been included in the New Japan Industry Hall in order to identify any shortcomings on the part of Japanese goods and to improve Japanese standards.[70]

The America Fair was part of what the *Asahi Shimbun* called the "*Hakurankai no rasshu jidai*" ("Age of the Exposition"). With the revival of private foreign trade and export promotion, the newspaper noted that Japan had been invited to display goods at eleven overseas fairs and expositions. In addition, there were many exhibitions being mounted throughout Japan. The great rush to do so began in 1949 with eight domestic fairs being held: a trade fair in Yokohama, a peace expo in Nagano and fairs in Sendai, Fukushima, Matsuyama, Takamatsu, Okayama and Tsuyama.[71]

The Kobe Fair

In 1950, a total of ten domestic fairs were scheduled to be held with the direct support of the Ministry of International Trade and Industry (MITI). Notable among these were the America Fair in Nishinomiya and the Japan Trade and Industry Fair also known as the Kobe Fair 1950 which ran for three months from 15 March through to 15 June. The main site of the Kobe fair was Ōji Park which directly overlooked the port. The secondary site was Minatogawa Park. In contrast to the America Fair which was aimed at selling the American dream to the Japanese people, the Kobe Fair sought to showcase the progress that the Japanese had made in foreign trade and industry.[72] Allied Occupation personnel also facilitated the Kobe

Fair in terms of working out the scheme and what the fair would entail, as well as helping to collect exhibits on display. In this way, GHQ, SCAP helped to show what America had to offer at Nishinomiya and in nearby Kobe, they could help project an image of the new Japan to the Japanese people and foreign visitors.[73]

The main aim of the Kobe Fair was to aid Japan's economic reconstruction and thereby improve the lives of the Japanese people. The organizers turned to architects and artists associated with the Seisakuha Kyōkai (New Production Group Association) for design ideas about how to translate this into an exposition. Members of the group conceived the theme in terms of sub-themes, all of which would form part of a symphony. Visitors to the fair would be greeted by a 70-metre-long wall at the entrance to the Ōji Park venue, designed by members of the Shin Seisakuha Association. The decorative relief was entitled "Trade, Industry and Peace" and symbolized the focus of the fair and indeed post-war Japan.

The first building was called the "Hall of Overture" where the message was that Japan had to regain its economic independence before 1953 and explained how it would go about doing that. Building on that message, other buildings represented five "movements" of that symphony with pavilions devoted to natural resources, the world, production, international trade and culture.[74] The building representing the world showed foreign exhibits which included the latest machines, household utensils, clothing, art works and books. A diagram illustrating the organization of the United Nations and its functions was displayed prominently on one wall.

In terms of the promotion of Japanese products, some prefectures built their own pavilions which encircled the main thematic pavilions while others showcased their local products in the two Production Halls. Kobe Steel, Kawasaki Heavy Industries and the Central Japan Heavy Industry companies were among the large companies that erected their own pavilions.[75] The International Trade Halls had sections devoted to land and marine transportation which showed how Japan was linked to the rest of the world. The halls provided detail information on the state of Japan's foreign trade.

It was in the Culture Hall where visitors could obtain a sensory experience about the new Japan. They could view exhibits on the latest forms of entertainment that include films and radio, television, books and theatre as well as learn about the latest fashions and how they could improve their everyday life by the way they dressed and did their hair, what they ate, how they prepared their food to the type of housing they lived in. There were sections in the Culture Hall devoted to work and recreation. Thanks to exhibits provided by the CIE Library, visitors were able to view photographs and models about the life of workers and it was notable that women's occupations were also displayed in line with Occupation efforts to promote the participation of women in the workforce.[76]

It was also in the Culture Hall where visitors could engage with and learn more about science. Displays on geophysics, geology and geography explained Japan's place in the world, and it was here that profiles of Nobel Prize winners were displayed. Japan's Yukawa Hideki had recently won the 1949 Nobel Prize in physics for his prediction of the existence of mesons as part of his theoretical work on nuclear forces. And it was here in the Culture Hall that visitors would learn about features of the Atomic Age.[77]

In the "finale" section of the fair, there were exhibits by foreign missions and embassies which gave the fair a further international flavour and promoted the idea "that the political organization of the world in an atomic age should be world government and a world without oppression and fear of war."[78] As part of the finale display, the Seisakuha Kyōkai member, Kikuchi Kazuo who was a distinguished professor at Tokyo University of the Arts, produced his iconic *Smiling Goddess* sculpture which consisted of a female figure at its centre, standing high on a plinth, holding up a torch with one arm. Around the female goddess swirled two large metal rings which suggested the orbits of electrons. In this way, Kikuchi presented a highly gendered symbol of a benign atom, an atom for peace.[79]

Kikuchi would go on to make the iconic sculpture *Statue for the Children of the Atomic Bomb* (also known as *The Children's Peace Monument*) which was erected several years later at the Hiroshima Peace Memorial Park to commemorate the life of the young girl Sasaki Sadako who died of leukemia due to the dropping of the atomic bomb on Hiroshima.[80] At the top of what appears to be the stylized shell of an atomic bomb stands a young female figure representing Sadako who with outstretched arms supports a large wire crane above her head. Below Sadako, flanking each side of the shell are figures of a boy and a girl to represent the thousands of other children who fell victim to the bomb. Inside the shell itself has hung a bronze crane suspended from a bell, both of which were donated by the Nobel Prize-winning physicist Yukawa Hideki.

What is fascinating to see is how Kikuchi's sculptures contributed to discourses about nuclear power. As the very first nation to fall victim to the atomic bomb in wartime, some Japanese and Americans thought that

Japan should embrace atoms for peace. At the same time, the Hiroshima experience allowed Japan to promote a discourse about the need for universal peace. What emerged was the image of Japan as a peace-loving nation that fights against nuclear weapons. As Roni Sarig suggests, Sadako's story was at first promoted by the Austrian journalist Robert Jungk and the German author Karl Bruckner before being taken up by the Japanese themselves. Sabine Frühstück has noted the importance of the bombing of Hiroshima and Nagasaki in the production of the victim culture that has characterized post-war Japan. Kikuchi's sculpture at Hiroshima contributed to this culture by highlighting Sadako as a symbol of suffering and the need for peace.[81] These discourses are therefore not entirely of Japanese making but can be viewed as co-productions which ultimately became a part of Japanese national discourses about their past and future.[82] Exhibitions were one way of spreading these narratives to both Japanese and foreign audiences.

With all the fairs scheduled to be held in Japan in 1950, the Japan Travel Bureau (JTB) was optimistic and estimated that there would be 23,589 visitors to Japan that year who would spend the equivalent of US$11,555,125. This represented one half the total of the pre-war peak tourist years of 1937 and 1938. Matsumura Nobuo, chief of the Foreign Tourist Section, JTB, considered one of the biggest obstacles was the lack of good, foreign-style hotels, even in Tokyo.[83] To solve this problem in Kobe, an Inland Sea liner was docked at the Central Pier at Kobe to help accommodate visitors from out-of-town. It provided a temporary "ship hotel" experience for some 700 visitors who visited the fair by connecting bus.[84] Other cities were also keen to attract visitors.

Prominent among the other fairs held in Japan in 1950 was an Industrial Culture Expo which ran in Tokyo from 15 March to the end of May, and a Hokkaidō Development Expo held in Asahikawa from 15 July to 23 August. The *Asahi Shimbun* attributed this rush to hold fairs as being driven not only by a desire to attract tourists and foreign buyers but also as a way of building infrastructure—venues that could be later used as public halls.[85] For example, after the Kobe Fair, the Resources Hall, the largest building on the main site, was set to become a municipal recreation centre that could accommodate 10,000 people and include a 50-metre swimming pool, amphitheatre, auditorium and gymnasium.[86] In this way, expositions were a crucial way of rebuilding Japan, especially for cities such as Kobe, 70 per cent of which had been destroyed during the war. In the five years after Japan's defeat, approximately 60 per cent of the city had been reconstructed and the Kobe Fair facilitated that, even though the fair itself was not a financial success. It did, however, attract 1.8 million visitors as well as delivering much-valued infrastructure, so from the viewpoint of the organizers Hyōgo prefecture and the city of Kobe, the results justified the expense.[87]

The Democratization of Japan Exhibition

Baker was also involved in the *Democratization of Japan* exhibition held at the San Francisco Memorial Opera House in September 1951 on the occasion of the conference where the peace treaty was signed between Japan and the Allied Powers. Flown to San Francisco with five members of her staff, the team mounted visual displays of the great political, social and economic changes that the Occupation had helped bring about. Delegates entering the conference venue would be able to see for themselves, in the outer hallway where the exhibition was located, the transformations that had occurred.[88]

Baker and her team showed considerable creativity in the way that they mounted the exhibition. Surviving photographs of the exhibition show Baker standing in front of the part of the display on fishing. Large photographs of men and women involved in the industry were mounted on a temporary wall, in front of which was positioned a basket and an actual fishing net draped diagonally across the display. Similarly, another display entitled "Women's Status Has Been Elevated" included not only photographs and explanatory text panels but also a bamboo basket with a flower arrangement to suggest depth.[89] These and other parts of the exhibition greeted delegates as they entered the Opera House. They provided visual evidence for President Harry S. Truman's claims in his speech at the opening of the Japanese Peace Treaty Conference:

> The occupation was designed by the wartime allies to prevent future Japanese aggression and to establish Japan as a peaceful and democratic country, prepared to return to the family of nations. The United States, as the principal occupying power, was given a special responsibility to carry out these objectives. It is our judgment that they have been achieved.[90]

Truman's speech on 4 September 1951 and subsequent days of the conference were televised by NBC throughout the USA.[91]

The exhibition supported Japanese Prime Minister Yoshida Shigeru's claim at the conference on 7 September that "The Japan of today is no longer the Japan of yesterday. We will not fail your expectations of us as a new nation, dedicated to peace, democracy and freedom."[92] *The New York Times* acknowledged that

> Many political reforms have been made during the Occupation. There is at least the façade of democracy: a bill of rights, universal suffrage, etc. But no one can be sure that a sovereign Japan will not gradually return to the old, authoritarian ways.[93]

It was almost as if the newspaper was referring to the images of Occupation reforms that greeted delegates in the hallway of the Opera House. Despite all the talk of peace by Truman, Yoshida and other speakers, *The New York Times* heralded the signing of the treaty as ushering in a new phase in the Cold War and what it termed "the struggle for Asia." As a result of the treaty, the West was seen as having been given an immediate advantage.

The newspaper also acknowledged that the absence of much of Asia at the conference gave Russians a propaganda weapon to criticize the proceedings.[94] Indeed, the Russian delegate Andrei A. Gromyko, Soviet Deputy Foreign Minister, issued a statement at a press conference on 8 September criticizing the American-British draft of the peace treaty as "clearing the path for a revival of Japanese militarism" and providing "a conversion of Japan into an American military base."[95] He claimed that

> it is not accidental that the preparation of a peace treaty with Japan was entrusted to such a seasoned warmonger as John Foster Dulles, who like an enterprising traveling salesman made a tour of capitals of certain countries imposing on the governments of those countries a draft peace treaty favoured by the United States.[96]

Despite Gromyko's complaints, the San Francisco Peace Treaty was signed that day on 8 September by forty-eight nations and the ceremony was broadcast by NBC over its radio and television networks. Under Article 3 of the Treaty, the USA retained provisional control of the Ryūkyū and Bonin Islands with a view to potential US administration under a United Nations trusteeship. The islands were ultimately returned to Japan. Japan and the USA also signed a separate security treaty that day.[97] While images of what the American-led Occupation had achieved in Japan and vision of

American diplomacy at work were important throughout the conference, there was a marked absence. General Douglas MacArthur who had been so instrumental in making the Peace Treaty Conference possible was not in attendance.[98] He had been relieved of his duties by Truman earlier that year on 11 April 1951.[99]

The Treaty came into effect on 28 April 1952, and with that, the Occupation officially came to an end. Determined to stay on in Tokyo, Baker transferred to the US Embassy where she became chief exhibition officer for the USIS/USIA and married Tokyo-based lawyer Thomas Blakemore in June 1954.[100] Chapters 4 and 5 examine some exhibitions that she worked on in that capacity.

Conclusion

The dropping of the atomic bomb on Hiroshima was a major turning point in Japanese history, but there were significant continuities if we look at pre-war and post-war Japan. Even before the end of the war, US media seemed to be absolving Emperor Hirohito of wartime responsibility and constructing a narrative that enabled him to stay in place as the nation's monarch. The Allied Occupation enabled Americans such as Frances Baker to renew their ties with Japan. Others such as Clark Goodman would build new links with Japan and meet key Japanese figures interested in atomic energy. What is striking in all this is the role of visual representations in mediating the US-Japan relationship. Fairs were highly visible ways by which the Japanese public could enjoy a taste of America, and build a new, more democratic Japan that looked to a future that was strongly aligned with the USA.

Notes

1. "Who Runs the Emperor," *Fortune* 29, no. 4 (Apr. 1944): 130–133, 270, 273, 274, 276, 279, 280, 283, 285, 286, esp. 131.
2. "Who Runs the Emperor," 131.
3. "Who Runs the Emperor," 133.
4. "What to do with Japan," 180–184, 289.
5. "Through the Looking Glass," advertisement, *Fortune* 29, no. 4 (April 1944): 288.
6. Jeffrey L. Meikle, *American Plastic: A Cultural History* (New Brunswick, New Jersey: Rutgers University Press, 1995), 176–177.

7. "Fortune's Wheel," *Fortune* 29, no. 4 (Apr. 1944): 2, 4, esp. 4.
8. Major Eric S. Fowler, *Japan's Imperial Institution and the U.S. Strategy to End World War II* (Fort Leavenworth, Kansas: US Army Command and General Staff College, 2012), 24–26.
9. Hugh Borton, *Spanning Japan's Modern Century: The Memoirs of Hugh Borton* (Lanham, Maryland: Lexington Books, 2002), 108–109.
10. Allison B. Gilmore, "'We Have Been Reborn': Japanese Prisoners and the Allied Propaganda War in the Southwest Pacific," *Pacific Historical Review* 64, no. 2 (May 1995): 195–215, esp. 210.
11. Memorandum to Director, OSS from Operations Officer, "Joint Anglo-American Outline Plan for Psychological Warfare Against Japan, CCS [Combined Chiefs of Staff] 539/4," July 8, 1944. Secret. Office of Strategic Services, Washington DC. CIA Electronic Reading Room.
12. PBS, *History* Detectives, Episode 902, Story 1 "WWII Leaflets," 29 June, 2011, transcript, http://www-tc.pbs.org/opb/historydetectives/static/media/transcripts/2011-06-29/902_WWIIleaflets.pdf
13. "Notes of the Interim Committee Meeting, May 31, 1945," 8. Top Secret. Miscellaneous Historical Documents Collection, no. 736, Harry S. Truman Library and Museum.
14. "Notes of the Interim Committee Meeting, May 31, 1945," 9.
15. "Notes of the Interim Committee Meeting, May 31, 1945," 9.
16. "Notes of the Interim Committee Meeting, May 31, 1945," 13.
17. "Notes of the Interim Committee Meeting, May 31, 1945," 14.
18. Memorandum from Arthur B. Compton to the Secretary of War, enclosing "Memorandum on 'Political and Social Problems,' from Members of the 'Metallurgical Laboratory' of the University of Chicago," 12 June, 1945, Secret. Enclosed Memorandum, 8–9. US National Archives and Records Administration, Record Group 77, Records of the Chief Engineers, Manhattan Engineering District, Harrison-Bundy files, folder no. 76. George Washington University National Security Archive (hereafter referred to as GWU NSA).
19. "Memorandum on 'Political and Social Problems,' from Members of the 'Metallurgical Laboratory' of the University of Chicago," 12 Jun. 1945, Secret, 10.
20. Memorandum by J. R. Oppenheimer, "Recommendations on the Immediate Use of Nuclear Weapons," 16 Jun. 1945, Top Secret. US National Archives, Record Group 77, Records of the Chief Engineers, Manhattan Engineering District, Harrison-Bundy files, folder no. 76. GWU NSA.
21. "The Job Before Us: The Defeat of Japan," *Fortune* 31, no. 6 (Jun. 1945): 113.

22. Hugh Borton, *Spanning Japan's Modern Century: The Memoirs of Hugh Borton* (Lanham, Maryland: Lexington Books, 2002), 148–149.
23. Memorandum from General L. R. Groves to the Chief of Staff, 6 Aug. 1945, 2. Top Secret. US National Archives, Record Group 77, Records of the Chief Engineers, Manhattan Engineering District, Top Secret Documents, File no. 5b. GWU NSA.
24. Cable CAX 51813 from USS Teton to Commander in Chief Army Forces Pacific Administration Manila Philippines. From Major General Thomas F. Farrell to Major General Leslie R. Groves, 10 Sept. 1945, 1, Secret. GWU NSA.
25. Cable CAX 51948 from Commander in Chief Army Forces Pacific Advance Yokohama Japan to Commander in Chief Army Forces Pacific Administration Manila Philippines, 14 Sept. 1945, 1, Secret. US National Archives, Record Group 77, Tinian Files, April–December 1945, box 17, envelope B. GWU NSA.
26. Michihiko Hachiya, *Hiroshima Diary: The Journal of a Japanese Physician August 6–September 30, 1945*, trans. and ed. Warner Wells (Chapel Hill: Univ. of North Carolina Press, 1955), 183–185.
27. "Text of Hirohito's Radio Rescript," *New York Times*, 15 Aug. 1945, 3.
28. John W. Dower, "Commentary: "Culture," Theory, and Practice in U.S.-Japan Relations," *Diplomatic History* 24, no. 3 (Summer 2000): 517–528, esp. 517; John W. Dower, *Embracing Defeat: Japan in the Wake of World War II* (New York: W.W. Norton and Co., 1999), 291.
29. GHQ, SCAP, CIE, "Memorandum for Imperial Japanese Government: Abolition of Governmental Sponsorship, Support, Perpetuation, Control, and Dissemination of State Shinto (*Kokka Shintō, Jinja Shintō*)," 15 Dec.1945. Reprinted in *Contemporary Religions in Japan* 7, no. 4 (Dec. 1966): 354–360.
30. "Our Policy in the Far East," *New York Times*, 17 Dec. 1945, 20.
31. Mark R. Mullins, "Religion in Occupied Japan: The Impact of SCAP's Policies on Shinto," in *Belief and Practice in Imperial Japan and Colonial Korea*, ed. Emily Anderson (Singapore: Palgrave Macmillan, 2017), 229–248, esp. 235.
32. Hiromi Mizuno, *Science for the Empire: Scientific Nationalism in Modern Japan* (Stanford: Stanford University Press, 2009), 2.
33. Barak Kushner, "Japan's War of Words with the World: WWII Propaganda in the International Arena," in *Routledge Handbook of Modern Japanese History*, eds. Sven Saaler and Christopher W.A. Szpilman (London: Routledge, 2018), 251–263.
34. See Morris Low, *Japan on Display: Photography and the Emperor* (New York: Routledge, 2006).

35. Leonard Weiss, "Atoms for Peace," *Bulletin of the Atomic Scientists* 59, no. 6 (Nov.-Dec. 2003): 34–44.
36. US Strategic Bombing Survey, *Japan's Struggle to End the War* (Chairman's Office, 1 Jul. 1946), 41. Elsey Papers, Harry S. Truman Administration. Japan, Surrender of, Aug. 1945. Truman Library and Museum. Also, see Gian Peri Gentile, "Advocacy or Assessment? The United States Strategic Bombing Survey of Germany and Japan," *Pacific Historical Review* 66, no. 1 (Feb. 1997): 53–79.
37. Dominic Kelly, "Ideology, Society, and the Origins of Nuclear Power in Japan," *East Asian Science, Technology and Society: An International Journal* 9 (2015): 47–64.
38. John C. Allred, William R. Stratton, Robley D. Evans, and M. Stanley Livingston, "Clark Goodman," *Physics Today* 37, no. 1 (1984): 96–97.
39. Clark Goodman, ed., *Science and Engineering of Nuclear Power*, 2 vols. (Cambridge, Mass.: Addison-Wesley Press, 1947–49).
40. Goodman, ed., *Science and Engineering of Nuclear Power*, vol. 2.
41. Clark Goodman, "Future Developments in Nuclear Energy," in *Science and Engineering of Nuclear Power*, vol. 2, ed. Clark Goodman (Cambridge, Mass.: Addison-Wesley Press, 1947–49), 275–296, esp. 275.
42. Goodman, "Future Developments in Nuclear Energy," esp. 275.
43. Goodman, "Future Developments in Nuclear Energy," esp. 275.
44. Civil Information and Education Section, GHQ, SCAP, *Mission and Accomplishments of the Occupation in the Civil Information and Education Fields*, 1 Jan. 1950, 1.
45. Shunya Yoshimi, trans. David Buist, "`America' as Desire and Violence: Americanization in Postwar Japan and Asia during the Cold War," *Inter-Asia Cultural Studies* 4, no. 3 (2003): 433–450, esp. 435.
46. Michiyo Morioka, *An American Artist in Tokyo: Frances Blakemore, 1906–1997* (Seattle: The Blakemore Foundation, 2007), 8, 86; for photographs of Baker and her colleagues at work as well as members of the public viewing the exhibits see Yokohama Kokusai Kankeishi Kenkyūkai, Yokohama Archives of History, *Zusetsu Don Buraun to Shōwa no Nihon: Korekushon de miru senji senryō seisaku* (*Don Brown and Showa Japan (with illustrations): Wartime and Occupation Policy as Seen in the Collection*) (Yokohama: Yūrindō, 2005), 89. Don Brown was Chief of the Information Division.
47. Civil Information and Education Section, *Mission and Accomplishments of the Occupation in the Civil Information and Education Fields*, 2.
48. Morioka, *An American Artist in Tokyo*, 86; Civil Information and Education Section, *Mission and Accomplishments of the Occupation in the Civil Information and Education Fields*, 10–11.

49. Shunya Yoshimi, "Consuming `America': From Symbol to System," in *Consumption in Asia: Lifestyles and Identities*, ed. Chua Beng-Huat (London: Routledge, 2000), 202–224, esp. 206.
50. Frances Baker, Exhibits Officer, Exhibits Branch, Information Division, Civil Information and Education Section, GHQ, SCAP, "Exhibits Branch Activities Report for Two Weeks Ending 15 November 1949," accessed 18 Aug. 2018, http://webarchives.tnm.jp/docs/cie/database/5325-17-13_en.html.
51. Baker, "Exhibits Branch Activities Report for Two Weeks Ending 15 November 1949."
52. Yuka Tsuchiya, "Imagined America in Occupied Japan: (Re-)Educational Films Shown by the U.S. Occupation Forces to the Japanese, 1948–1952," *The Japanese Journal of American Studies*, no. 13 (2002): 193–213; Tsuchiya, Yuka, "Amerika tai Nichi senryōgun `CIE Eiga': Kyōiku to puropaganda no kyōkai" ("US Occupation Forces in Japan's CIE Films: The Boundaries of Education and Propaganda"), *Ehime Daigaku Hōbungakubu ronshū, sōgō seisakugakka hen* (*Papers of the Ehime University Faculty of Law and Letters, General Policy Section*) 19 (2005): 27–54, esp. 48.
53. "Preview of American Fair," *Nippon Times*, 17 Mar. 1950, 3.
54. See the Asahi Shimbun photograph of her doing so: "America Fair Formally Opened," *Nippon Times*, 20 Mar. 1950, 1.
55. Morioka, *An American Artist in Tokyo*, 100.
56. *The America Fair* (Tokyo: Asahi Press, c. 1950), 32–33.
57. Morioka, *An American Artist in Tokyo*, 101; Iwamoto Shigeki, "Burondi (2): Sengo Nihon ni okeru Amerikanizeeshon" ("Blondie: Americanization in Japan after the Second World War, Part 2"), *Shakaigakubu kiyō* (*Bulletin of the School of Sociology*) 79 (March 1998): 147–160.
58. Morioka, *An American Artist in Tokyo*, 101; Iwamoto, "Burondi (2)"; "`America Fair' Being Sponsored by Asahi Shimbun," *Nippon Times*, 17 Mar. 1950, 3.
59. *The America Fair*, 95.
60. Iwamoto, "Burondi (2)": 154–155.
61. Asahi Shimbun and Nippon Times, "Filmization is Slated," *Nippon Times*, 23 Mar. 1950, 3.
62. "America Fair: Replicas of Present-Day U.S. Brought to Japan at Asahi Shimbun Exhibit," *Nippon Times*, 23 Mar. 1950, 3.
63. "Nugent Speaks at Fair Opening," *Nippon Times*, 23 Mar. 1950, 5.
64. "Nugent Speaks at Fair Opening."
65. "Nugent Speaks at Fair Opening."
66. "Nugent Speaks at Fair Opening."
67. "Nugent Speaks at Fair Opening."

68. Earnest Hoberecht, UP Manager for Japan, "American Bases in Japan," *Nippon Times*, 25 Mar. 1950.
69. "'Little America' in Japan," *Popular Mechanics* (Jun. 1950), 96–97.
70. Tadasu Hasebe, "Why the America Fair," excerpts from a speech given at the Opening Ceremony on 17 Mar. 1950, *The America Fair*.
71. "Hakurankai no rassshu jidai" ("Age of the Exposition"), *Asahi Shimbun*, 29 Jan. 1950, 3.
72. Tom Lambert, "Huge Kobe Fair to Open Mar. 15," *Nippon Times*, 6 Mar. 1950, 5.
73. Takeo Saito, "In Commemoration of the Kobe Fair," in Japan Foreign Trade and Industry Exposition Administration Bureau (ed.), *Nihon Bōeki Sangyō Hakurankai 'Kōbe Haku' kaishi, 1950* (*The Japan Foreign Trade and Industry Exposition Report, 1950*) (Kobe: The Exposition, 1951), 1–10.
74. Funabiki, Etsuko, "Funabiki Etsuko no ronbun kara" (From Etsuko Funabiki's thesis,") *Shinseisaku kaihō*, no. 52 (15 Dec., 2006): 7–8; "Huge Kobe Fair Opens March 15," *Nippon Times*, 12 Mar. 1950, 3. See also Funabiki, Etsuko and Umemiya, Hiromitsu, "The Planning of Japan Foreign Trade and Industry Exposition, The Kobe Fair 1950, by Architects of Shinseisaku," *Ningen kagaku kenkyū* (*Human Sciences Research*) 10, no. 1 (2002): 89–112.
75. "Trade Tour Fair Opens to Public," *Nippon Times*, 23 Mar. 1950, 4.
76. Saito, "In Commemoration of the Kobe Fair," 6.
77. Saito, "In Commemoration of the Kobe Fair," 6.
78. "Huge Kobe Fair Opens March 15," *Nippon Times*, 12 Mar. 1950, 3.
79. Japan Foreign Trade and Industry Exposition, *Nihon Bōeki Sangyō Hakurankai 'Kōbe Haku' kaishi, 1950* (*The Japan Foreign Trade and Industry Exposition Report, 1950*) (Kobe: The Exposition, 1951), 1.
80. For details of Kikuchi's involvement, see Masamoto Nasu, *Children of the Paper Crane: The Story of Sadako Sasaki and Her Struggle with the A-Bomb Disease*, trans. Elizabeth W. Baldwin, Steven L. Leeper and Kyoko Yoshida (Armonk, N.Y.: M.E. Sharpe, 1991), 180–181.
81. Sabine Frühstück, "'…And my heart screams': Children and the War of Emotions," in *Child's Play: Multi-Sensory Histories of Children and Childhood in Japan*, eds. Sabine Frühstück and Anne Walthall (Oakland, Calif.: University of California Press, 2017), 181–201, esp. 194.
82. Roni Sarig, "Sadako Sasaki and Anne Frank: Myths in Japanese and Israeli Memory of the Second World War," in *War and Militarism in Modern Japan: Issues of History and Identity*, ed. Guy Podoler (Folkestone, Kent: Global Oriental, 2009), 158–171.
83. "23,000 Tourists Expected in 1950," *Nippon Times*, 10 Apr. 1950, 3.
84. "Huge Kobe Fair Opens March 15," *Nippon Times*, 12 Mar. 1950, 3.

85. "Hakurankai no rassshu jidai" ("Age of the Exposition"), *Asahi Shimbun*, 29 Jan. 1950, 3.
86. "Trade Tour Fair Opens to Public," *Nippon Times*, 23 Mar. 1950, 4; "Huge Kobe Fair Opens March 15," *Nippon Times*, 12 Mar. 1950, 3.
87. Saito, "In Commemoration of the Kobe Fair," 1–10.
88. Morioka, *An American Artist in Tokyo*, 101.
89. Morioka, *An American Artist in Tokyo*, 101.
90. President Truman, "Statements by Truman, Robinson and Warren at the Opening of Conference," *New York Times*, 5 Sept., 1951, 6. See also US Government, Department of State, *Conference for the Conclusion and Signature of the Treaty of Peace with Japan; San Francisco, California, September 4–8, 1951; Record of Proceedings* (Washington D.C.: US Government Printing Office, 1951), 32.
91. "NBC to Cover Peace Signing," *Los Angeles Sentinel*, 6 Sept. 1951, A5. See also James Von Schilling, *The Magic Window: American Television, 1939–1953* (New York: Haworth Press, 2003), 172.
92. "Sovereign Japan," *New York Times*, 9 Sept. 1951, 1.
93. "Sovereign Japan."
94. "Sovereign Japan."
95. "Text of Gromyko's Statement on the Peace Treaty," *New York Times*, 9 Sept. 1951, 26–27, esp. 26.
96. "Text of Gromyko's Statement on the Peace Treaty," 27.
97. Kuniyoshi Tomoki, "San Francisco Peace Treaty (1951)," in *Japan at War: An Encyclopedia*, ed. Louis G. Perez (Santa Barbara, Calif.: ABC-CLIO, 2013), 359.
98. "Peace 10 Years Later," *Los Angeles Sentinel*, 13 Sept. 1951, A8.
99. "M'Arthur Ouster Stuns L.A.," *Los Angeles Sentinel*, 12 Apr. 1951, A1–A2.
100. Morioka, *An American Artist in Tokyo*, 101.

Bibliography

Allred, John C., William R. Stratton, Robley D. Evans, and M. Stanley Livingston, "Clark Goodman." *Physics Today* 37, no. 1 (1984): 96–97.

Asahi Shimbun, 29 Jan. 1950.

Baker, Frances. Exhibits Officer, Exhibits Branch, Information Division, Civil Information and Education Section, GHQ, SCAP, "Exhibits Branch Activities Report for Two Weeks Ending 15 November 1949." Accessed 18 Aug. 2018 http://webarchives.tnm.jp/docs/cie/database/5325-17-13_en.html.

Borton, Hugh *Spanning Japan's Modern Century: The Memoirs of Hugh Borton*. Lanham, Maryland: Lexington Books, 2002.

Civil Information and Education Section, GHQ, SCAP, *Mission and Accomplishments of the Occupation in the Civil Information and Education Fields.* 1 Jan., 1950.

Dower, John W. "Commentary: "Culture," Theory, and Practice in U.S.-Japan Relations." *Diplomatic History* 24, no. 3 (Summer 2000): 517–528

Dower, John W. *Embracing Defeat: Japan in the Wake of World War II.* New York: W.W. Norton and Co., 1999.

Elsey Papers, Harry S. Truman Administration. Japan, Surrender of, August, 1945. Truman Library and Museum.

"Fortune's Wheel." *Fortune* 29, no. 4 (Apr. 1944): 2, 4.

Fowler, Major Eric S. *Japan's Imperial Institution and the U.S. Strategy to End World War II.* Fort Leavenworth, Kansas: US Army Command and General Staff College, 2012.

Funabiki, Etsuko, "Funabiki Etsuko no ronbun kara" (From Etsuko Funabiki's thesis.") *Shinseisaku kaihō,* no. 52 (15 Dec. 2006): 7–8

Frühstück, Sabine. "'...And my heart screams': Children and the War of Emotions." In *Child's Play: Multi-Sensory Histories of Children and Childhood in Japan*, edited by Sabine Frühstück and Anne Walthall, 181–201. Oakland, Calif.: University of California Press, 2017.

Funabiki, Etsuko and Umemiya, Hiromitsu, "The Planning of Japan Foreign Trade and Industry Exposition, The Kobe Fair 1950, by Architects of Shinseisaku."*Ningen kagaku kenkyū* (*Human Sciences Research*) 10, no. 1 (2002): 89–112.

Gentile, Gian Peri. "Advocacy or Assessment? The United States Strategic Bombing Survey of Germany and Japan." *Pacific Historical Review* 66, no. 1 (Feb. 1997): 53–79.

George Washington University, National Security Archive

GHQ, SCAP, CIE, "Memorandum for Imperial Japanese Government: Abolition of Governmental Sponsorship, Support, Perpetuation, Control, and Dissemination of State Shinto (*Kokka Shintō, Jinja Shintō*)." 15 Dec. 1945. Reprinted in *Contemporary Religions in Japan* 7, no. 4 (Dec. 1966): 354–360.

Gilmore, Allison B. "'We Have Been Reborn': Japanese Prisoners and the Allied Propaganda War in the Southwest Pacific." *Pacific Historical Review* 64, no. 2 (May 1995): 195–215.

Goodman, Clark, ed. *Science and Engineering of Nuclear Power*, 2 vols. Cambridge, Mass.: Addison-Wesley Press, 1947–49.

Goodman, Clark. "Future Developments in Nuclear Energy." In *Science and Engineering of Nuclear Power*, vol. 2, edited by Clark Goodman, 275–296. Cambridge, Mass.: Addison-Wesley Press, 1949.

Hachiya, Michihiko. *Hiroshima Diary: The Journal of a Japanese Physician August 6-September 30, 1945*, trans. and ed. Warner Wells. Chapel Hill: Univ. of North Carolina Press, 1955.

Hasebe, Tadasu. "Why the America Fair." Excerpts from a speech given at the Opening Ceremony on 17 Mar. 1950. *The America Fair*. Tokyo: Asahi Press, c. 1950.

Iwamoto, Shigeki. "Burondi (2): Sengo Nihon ni okeru Amerikanizeeshon" ("Blondie: Americanization in Japan after the Second World War, Part 2"). *Shakaigakubu kiyō* (*Bulletin of the School of Sociology*) 79 (Mar. 1998): 147–160.

Japan Foreign Trade and Industry Exposition. *Nihon Bōeki Sangyō Hakurankai 'Kōbe Haku' kaishi, 1950* (*The Japan Foreign Trade and Industry Exposition Report, 1950*). Kobe: The Exposition, 1951.

Kelly, Dominic. "Ideology, Society, and the Origins of Nuclear Power in Japan." *East Asian Science, Technology and Society: An International Journal* 9 (2015): 47–64.

Kuniyoshi Tomoki. "San Francisco Peace Treaty (1951)." In *Japan at War: An Encyclopedia*, edited by Louis G. Perez, 359. Santa Barbara, Calif.: ABC-CLIO, 2013.

Kushner, Barak. "Japan's War of Words with the World: WWII Propaganda in the International Arena." In *Routledge Handbook of Modern Japanese History*, edited by Sven Saaler and Christopher W.A. Szpilman, 251–263. London: Routledge, 2018.

"'Little America' in Japan." *Popular Mechanics* (June 1950), 96–97.

Los Angeles Sentinel, 12 Apr. 1951; 6 Sept. 1951. 13 Sept. 1951.

Low, Morris. *Japan on Display: Photography and the Emperor*. New York: Routledge, 2006.

Meikle, Jeffrey L. *American Plastic: A Cultural History*. New Brunswick, New Jersey: Rutgers University Press, 1995.

Miscellaneous Historical Documents Collection, Harry S. Truman Library and Museum.

Mizuno, Hiromi. *Science for the Empire: Scientific Nationalism in Modern Japan*. Stanford: Stanford University Press, 2009.

Morioka, Michiyo. *An American Artist in Tokyo: Frances Blakemore, 1906–1997*. Seattle: The Blakemore Foundation, 2007.

Mullins, Mark R. "Religion in Occupied Japan: The Impact of SCAP's Policies on Shinto." In *Belief and Practice in Imperial Japan and Colonial Korea*, edited by Emily Anderson, 229–248. Singapore: Palgrave Macmillan, 2017.

Nasu, Masamoto. *Children of the Paper Crane: The Story of Sadako Sasaki and Her Struggle with the A-Bomb Disease*, translated by Elizabeth W. Baldwin, Steven L. Leeper and Kyoko Yoshida. Armonk, N.Y.: M.E. Sharpe, 1991.

New York Times, 15 Aug. 1945, 17 Dec. 1945, 5–9 Sept. 1951.

Nippon Times, 6–25 Mar., 10 Apr. 1950.

Office of Strategic Services, Washington DC records. CIA Electronic Reading Room.

PBS, *History* Detectives, Episode 902, Story 1 "WWII Leaflets." 29 Jun. 2011, transcript, http://www-tc.pbs.org/opb/historydetectives/static/media/transcripts/2011-06-29/902_WWIIleaflets.pdf

Saito, Takeo. "In Commemoration of the Kobe Fair." In *Nihon Bōeki Sangyō Hakurankai 'Kōbe Haku' kaishi, 1950* (*The Japan Foreign Trade and Industry Exposition Report, 1950*), edited by Japan Foreign Trade and Industry Exposition Administration Bureau, 1–10. Kobe: The Exposition, 1951.

Sarig, Roni. "Sadako Sasaki and Anne Frank: Myths in Japanese and Israeli Memory of the Second World War." In *War and Militarism in Modern Japan: Issues of History and Identity*, edited by Guy Podoler, 158–171. Folkestone, Kent: Global Oriental, 2009.

The America Fair. Tokyo: Asahi Press, c. 1950.

"The Job Before Us: The Defeat of Japan." *Fortune* 31, no. 6 (June 1945): 113.

"Through the Looking Glass." Advertisement, *Fortune* 29, no. 4 (Apr. 1944): 288.

Tsuchiya, Yuka. "Imagined America in Occupied Japan: (Re-) Educational Films Shown by the U.S. Occupation Forces to the Japanese, 1948–1952." *The Japanese Journal of American Studies*, no. 13 (2002): 193–213.

Tsuchiya, Yuka, "Amerika tai Nichi senryōgun 'CIE Eiga': Kyōiku to puropaganda no kyōkai" ("US Occupation Forces in Japan's CIE Films: The Boundaries of Education and Propaganda.") *Ehime Daigaku Hōbungakubu ronshū, sōgō seisakugakka hen* (*Papers of the Ehime University Faculty of Law and Letters, General Policy Section*) 19 (2005): 27–54.

US Government, Department of State. *Conference for the Conclusion and Signature of the Treaty of Peace with Japan; San Francisco, California, September 4–8, 1951; Record of Proceedings*. Washington D.C.: US Government Printing Office, 1951.

US National Archives and Records Administration. Record Group 77, Records of the Chief Engineers, Manhattan Engineering District.

Von Schilling, James. *The Magic Window: American Television, 1939–1953*. New York: Haworth Press, 2003.

Weiss, Leonard. "Atoms for Peace." *Bulletin of the Atomic Scientists* 59, no. 6 (Nov.-Dec. 2003): 34–44.

"What to do with Japan." *Fortune* 29, no. 4 (Apr. 1944): 180–184, 289.

"Who Runs the Emperor." *Fortune* 29, no. 4 (Apr. 1944): 130–133, 270, 273, 274, 276, 279, 280, 283, 285, 286.

Yokohama Kokusai Kankeishi Kenkyūkai, Yokohama Archives of History. *Zusetsu Don Buraun to Shōwa no Nihon: Korekushon de miru senji senryō seisaku* (*Don Brown and Showa Japan (with illustrations): Wartime and Occupation Policy as Seen in the Collection*). Yokohama: Yūrindō, 2005.

Yoshimi, Shunya. Trans. by David Buist. "`America' as Desire and Violence: Americanization in Postwar Japan and Asia during the Cold War." *Inter-Asia Cultural Studies* 4, no. 3 (2003): 433–450.
Yoshimi, Shunya. "Consuming `America': From Symbol to System." In *Consumption in Asia: Lifestyles and Identities*, edited by Chua Beng-Huat, 202–224. London: Routledge, 2000.

CHAPTER 3

Picturing Hiroshima

Introduction

In turning our attention to the importance of visualization, we can identify historical figures who played a hitherto neglected role in the history of US-Japan relations, and ultimately in how the Japanese people would see nuclear power. We have previously noted the important role of Frances Baker. In this chapter, we turn our attention to an artist-activist named Akamatsu Toshiko. Like Baker, her skills were of use in promoting wartime propaganda but after the war, Akamatsu would join forces with her husband Maruki Iri and produce what are known as the "Hiroshima panels" that served to remind the Japanese people, at a time of censorship, what had occurred at Hiroshima and Nagasaki.

Akamatsu Toshiko and Wartime Propaganda

Akamatsu (1912–2000) is one notable example of what we could call an art activist. Her life story shows how a combination of very real talent in the visual arts, life experience, courage and fortuitous circumstances enabled an independent-minded woman to take political action through the production of provocative images that would change perceptions of atomic energy. Although born in Hokkaido, she commenced her study of oil painting in 1929 in Tokyo at Joshi Bijutsu Senmon Gakkō (Women's School of Art), a private art school established in 1900 by the female

M. Low, *Visualizing Nuclear Power in Japan*, Palgrave Studies in the History of Science and Technology,
https://doi.org/10.1007/978-3-030-47198-9_3

educator Yokoi Tamako who sought to provide a pathway for young women to become independent through art-related activities.

Joshibi, as it was referred to, was one of the few institutions of higher education in Tokyo available to women who sought to study art beyond high school.[1] At Joshibi, Akamatsu specialized in Western-style painting.[2] Japanese women had in the past been discouraged from studying oil painting. Nihonga (Japanese-style painting) was deemed to be more traditional and more feminine. Akamatsu chose otherwise. She helped support herself while there by drawing portraits of people at Ueno Park in Tokyo. On her graduation from Joshibi in 1933, Akamatsu became a substitute teacher at an elementary school in Ichikawa city, Chiba prefecture which was close to Tokyo.[3] Meanwhile, she continued to paint and exhibit. In 1937, in a further expression of her independence, she travelled to Moscow to work for a year as a nanny and home tutor for the five children of Yuhashi Shigetō who had been appointed to the diplomatic post of First Interpreter at the Japanese embassy there.[4]

In July 1937, the Second Sino-Japanese War broke out with the Marco Polo Bridge incident which led to a full-scale conflict between China and Japan. There was increasing suppression of left-wing thought and people suspected of being Communist sympathizers were sometimes detained and jailed. One year later, the National Mobilization Law was passed by the Japanese Diet. It was against this background that Akamatsu returned to Tokyo in 1938. Thanks to the considerable savings that had been accrued due to her year in Moscow, she was able to rent a home and studio in Tokyo in the Nagasaki area of Toshima ward. Located near Ikebukuro Station, the area became known as Ikebukuro Montparnasse, as it was Tokyo's version of the famous artistic hub in Paris. In the 1930s and 1940s, Toshima ward consisted of clusters of such rented houses with attached studios. It is estimated that over five hundred artists lived in such accommodation. Akamatsu lived in the community with the grand name of Sakuragaoka Parthenon. Some sixty artists lived there, including Kitagawa Tamiji who lived there from 1937 to 1943. Typically, the artists were unmarried as the actual living quarters were only 10–15 square metres in area. The houses had a sink and toilet but no bath, so residents would have to use the public bathhouse.[5] But the houses did come with studios which were a more generous 50 square metres.[6]

In March 1939, Akamatsu exhibited her sketches of Moscow at the Kinokuniya Gallery in the Ginza district of Tokyo, having obtained approval in advance to hold the exhibition from the Tsukiji police station.

She sought to assuage concerns that the exhibition would be promoting political propaganda. She explained to the authorities that she had merely accompanied a diplomatic family to Moscow and that she was planning to show sketches that were completed while there.[7]

In September 1939, Hitler invaded Poland and France and Great Britain declared war on Germany. That very same month, Akamatsu's work was selected for inclusion in the 26th Nika Association exhibition, an annual exhibition of Western-style art. The association of progressive Japanese artists had originally been established around the time of the outbreak of the previous world war to provide artists with an alternative to exhibiting at government-sponsored exhibitions. Kitagawa was among the artists who regularly showed their work at the Nika exhibition. Around that time, she also encountered the work of Maruki Iri, a self-taught artist who produced avant-garde ink paintings.

Akamatsu was subsequently approached by the military to accompany the male novelist Niwa Fumio and travel to the South Pacific on a warship as an official war artist.[8] Such positions were highly coveted by many artists and it was highly unusual for female artists to be given the opportunity.[9] It was Western-style painting that was preferred for propagandistic depictions of Japanese soldiers at war.[10] Akamatsu was able to obtain a medical certificate for some feigned illness from a doctor whom she knew and was able to avoid such wartime collaboration.

In October 1939, Akamatsu went to the movies in Tokyo and saw newsreel footage of the island of Yap with her lover at the time, artist Yamamoto Naotake. They both dreamt of visiting what appeared on the screen to be a tropical paradise but after being jilted by Yamamoto some two months later, Akamatsu decided to start her life anew and left on a boat from the port of Yokohama on 19 January 1940, bound for Palau in Micronesia where she would spend some six months, living firstly on Palau and then in late April on Yap. On her way to Palau, on 22 January 1940, the boat passed the island of Tinian from where the Enola Gay would take off with the atomic bomb bound for Hiroshima in August 1945. Little did she know that what would subsequently occur in the Pacific would come to define her work for the rest of her life.[11]

A Japanese colonial government (Nan'yō-chō) had been established in Micronesia since World War I and, by the 1930s, employed some 950 officials to administer the islands. Japanese immigrants tended to settle in the larger islands such as Saipan, Palau and Ponape. In the 1930s, the Japanese colonial scholar Yanaihara Tadao visited the islands. Yanaihara, a

Christian, was concerned for the well-being of the Micronesians. The colonial government tended to see them as an economic burden, part of a primitive culture that was inherently childish and lazy. Yanaihara advocated their complete assimilation into Japanese modernity. Without such action, there was concern that the Micronesians would die away[12] and be replaced by Japanese immigrants. Micronesia would ultimately be ruled by Japan for three decades.[13] As historian Mark R. Peattie writes, the islands never became the setting for a major piece of fiction but it is noteworthy that they provide the background for some children's books that Akamatsu Toshiko (later known as Maruki Toshi) would illustrate during the war.

Soon after her arrival at Palau, Akamatsu met the Japanese artist Hijikata Hisakatsu, a long-term resident who was employed part-time by the South Seas colonial government (Nan'yō-chō) to help visitors like her to settle in and feel welcome. While there, Akamatsu made many sketches of the local people at a time when increasing numbers of Japanese were visiting and settling in the islands. Thanks to Hijikata, she was able to exhibit some of her work at the government staff club, the Shōnan Club on Palau.[14] One photo taken during her subsequent month-long visit to Yap shows her dressed in a grass skirt posing alongside a local indigenous girl, both beaming and bare-breasted.[15] It was as if she was part of footage that she had seen with Yamamoto! On her return to Japan, Akamatsu exhibited her sketches and paintings at the Kinokuniya Gallery in the Ginza in September 1940. Maruki Iri saw those works and asked to meet her.[16]

Despite a blossoming relationship, Akamatsu returned to Moscow for a second stint when the diplomat Nishi Haruhiko was appointed Minister Plenipotentiary to the Soviet Union in September 1940. He had previously been Counsellor at the embassy. This time, she worked as a tutor for Nishi's daughter, Miyoko, in Moscow from January 1941 until around May when the Nishi family returned to Japan after there was a falling out between Nishi and the Japanese ambassador to the Soviet Union, Lieutenant General Tatekawa Yoshitsugu.[17] Tatekawa had been appointed by the Foreign Minister Matsuoka Yōsuke. Akamatsu had the opportunity to meet both men at a celebratory party after they had signed the Soviet-Japanese Neutrality Pact, on behalf of Japan, in April 1941. The Nishi family's time in Moscow came to an early end soon after the signing of the non-aggression pact and just prior to the outbreak of the German-Soviet War in June 1941. Safely back in Japan, Mrs. Nishi served as the formal intermediary at the wedding ceremony of Akamatsu and Maruki Iri in

July[18] and not long after in October 1941, Nishi was appointed Vice-Minister for Foreign Affairs.[19]

In December 1941, Japanese naval and air forces based in Micronesia launched attacks on Pearl Harbor. The American artist Frances Baker (known as Frances Blakemore after her marriage) had just left Japan for Honolulu, prior to the outbreak of the Pacific War. During the war, she worked as chief illustrator in the art department at the Office of War Information, Honolulu, producing propaganda leaflets targeting the Japanese whom she had gotten to know so well. Akamatsu was meanwhile busy illustrating patriotic children's books, back in Japan, to help inspire the younger generation to contribute in their way to the war effort.[20]

Akamatsu had become interested in book illustration through a study group she had joined. The group included artists who lived nearby such as Kitagawa Tamiji and Terada Takeo, both of whom had lived and studied in the USA. Art educator Kubo Sadajirō also joined the group with a view to improving the quality of wartime children's books and publishing some titles. In late 1941, Kitagawa and Kubo launched a small publishing company Kodomo Bunkasha in Kitagawa's home.[21] It was hoped that such work would generate some income at a time when the market for art works was not strong, but wartime restrictions on paper severely limited the company's ability to print the books. So Akamatsu turned elsewhere.

Akamatsu herself would illustrate some seven children's books during the war, most of which were set in the South Pacific and published by bigger publishers. They included *Yashi no mino tabi* (*The Travels of a Palm Tree Fruit*) (1942). Although the author of the book is given as Maruyama Kaoru, Akamatsu actually wrote the text and Maruyama, a known children's literature author, added his hand to it.[22] The front cover of the book shows a cute native boy's face in amongst the fruit and fronds of a palm tree. And on the back we see the curled up figure of the boy with the fruit on his head to suggest that the fruit and the boy are one.[23] The fruit-boy is determined to become a fine tree and drifts off on the sea waters, only to be picked up by a Japanese warship and turned into a brush that is used to polish the deck. The fruit-boy is delighted that he can be of use to Japan. In this way, the little fruit-boy from the colonies does his bit to assist the Japanese empire and the war effort. The book, published by the Teikoku Kyōiku Shuppanbu (Imperial Education Press) in Tokyo, proved popular, selling some 10,000 copies in the first printing. It was reprinted the following year and it is estimated that around 5000 copies were sold.[24]

This was a major success for Akamatsu, given that she was responsible for so much of its content. Later it would come to be seen as a liability.

Akamatsu also illustrated a second book that was released by the well-known educational publisher Shōgakkan around the same time, Tsuchiya Yukio's *Yashi no ki no shita* (*Under a Palm Tree*) (1942).[25] The book takes the form of the journal of a boy who travels to Singapore in time for its surrender to the Japanese and its name change to "Shōnantō." He reports his delight on the fall of Singapore and, as the cover depicts, the raising of the Japanese flag above buildings throughout the city.[26] The book illustrations show the highly multicultural nature of the population and, in one night scene, we see what appears to be a sweet Japanese boy in a summer kimono bringing a clock (read "modernity") to a local inhabitant, standing under a palm tree in the moonlight.[27]

In the Aftermath of the Bombing of Hiroshima and Nagasaki

Akamatsu and Maruki Iri had married in 1941. He was the eldest son of a farming family in a village which is now part of Hiroshima city. Three days after the atomic bomb had been dropped on Hiroshima, Maruki travelled from Urawa in Saitama prefecture where he had been evacuated to, to Hiroshima to check on the welfare of his family. Akamatsu followed a week after. What they saw and heard about from atomic bomb survivors left a lasting impression on both of them. Sadly, Maruki Iri's father would soon die as a result of his injuries, but his mother Maruki Suma and his younger sister Daidō Ayako (known as Aya for short) would survive and later produce art works of their own.[28] The visit to Hiroshima would impact on Maruki Iri and Akamatsu's health as well, with Akamatsu suffering from residual radiation sickness in subsequent years.

Akamatsu and Maruki returned to Urawa in September 1945, in time for the beginning of the American-led Allied Occupation. The rhetoric was peace and reconstruction. Young Japanese were encouraged to be hopeful about the future and it was not a time to reflect on what they had witnessed at Hiroshima. The words "peace" and "reconstruction" were the keywords of the period and it would take them three years to come to a mutual decision that it was time to paint what they had seen and heard about in Hiroshima at the end of the war.[29]

The Marukis and the Allied Occupation

During the Occupation, the American Frances Baker returned to Japan and helped shape Japanese attitudes through the production of visual materials and exhibits to promote the Occupation's goals and campaigns. What she and her colleagues produced served to influence what the Japanese saw and thought, and helped manage perceptions of the USA. It was also through exclusion of material, most notably references to the atomic bomb and discussion and critique of the role of the USA in first using nuclear weapons against Japan, that it would seek to shape public discourse.

At a time of censorship, Akamatsu and Maruki were among the first to convey to Japanese throughout the country of what had taken place at Hiroshima, Nagasaki and elsewhere through their series of fifteen large-scale, jointly produced, Chinese ink on Japanese paper paintings often referred to as *genbaku no* zu (atomic bomb paintings) or Hiroshima panels which they started painting from the late 1940s. Each panel is approximately 1.8 by 7.2 metres (6 feet by almost 24 feet)[30] in size and made up of eight sections to make them more portable. The Civil Intelligence Section, General Headquarters, Supreme Commander for the Allied Powers reported on exhibitions of this series of paintings which they regarded as propaganda instruments for Communist-front organizations such as the Japan Peace Protection Committee (Heiwa Yōgo Nippon Iinkai) which they deemed to be the Japanese counterpart of the World Peace Council that was founded in 1950 and promoted by the Soviet Union. The artists, likewise, were considered by Civil Intelligence to be Communists and they had indeed apparently joined after returning to Tokyo from Hiroshima.[31]

The first of their large murals was entitled *August Sixth* in an attempt to avoid potential censorship. It was shown at the Third Exhibition of Japanese Independent Artists held at the Tokyo Metropolitan Art Museum in Ueno Park in February 1950 under the sponsorship of the Nihon Bijutsu Kai (Japan Art Association) which both Akamatsu and Maruki had joined as founding members. The association was also deemed by Civil Intelligence to be a Communist-front organization, and effectively became part of the cultural section of the Japanese Communist Party.[32] There were showings of the painting which later came to be known as *No. 1 Ghosts* at the Maruzen Gallery in Tokyo in March 1950. The Hiroshima panel paintings grew in number. The Marukis completed *No. 2 Fire* and *No. 3*

Water in August 1950 and exhibited the three panels at Maruzen Gallery and the Mitsukoshi Department Store in the Ginza district of Tokyo.[33]

The paintings subsequently toured the country under the auspices of the Japan Peace Protection Committee (JPPC), labour unions and other organizations. The tour was seen as part of the Japanese Communist Party's campaign to discredit the USA. As Civil Intelligence noted, the exhibitions were accompanied by lectures and discussions as part of the Party's so-called peace offensive. In this way, a performative element was added to displays of the atomic bomb paintings to make each showing an interactive and memorable experience for all who attended.[34]

The tour involved considerable organization at the local level. At some venues, admission was free and at others, people paid 10 yen to defray costs, depending on the availability of sponsorship. Handbills and posters were used to advertise the exhibitions at each location and the postcard-size photographic reproductions of the paintings were sold at venues. The JPPC usually paid the train fares of Akamatsu and Maruki but local hosts provided accommodation and the cost of the venue. Any profits were split three ways between the artists, the JPPC and an organization called the Overall Peace Patriotic Movement Council. At some venues, local branches of the Democratic Scientists Association (known as "Minka" for short) sponsored the exhibition.[35]

Civil Intelligence reported that one of the most effective features of the exhibit was a small pocket-size booklet called ***Pikadon*** (***Flash-boom***) which consisted of black and white reproductions of pen and ink line drawings. Edited by the defunct Heiwa o Mamoru Kai (Peace Preservation Society)[36] and published by the Potsdam Book Store in Tokyo, the book was sold at exhibition venues for 30 yen. The cover of ***Pikadon*** shows an old grandmother who, in the book, relates the impact of the atomic bomb on her family and friends. In this way, the book was seen as a way of extending the message of the exhibition to the broader community, including to those unable to attend the exhibition in person. Many of those who saw the exhibition were students, office workers and the intelligentsia. The exhibition is said to have left a particularly strong impression on the female students who attended.[37]

On 14 May 1951, the week-long Kyoto University Spring Cultural Festival began. At the festival, science and medical students created exhibits exploring aspects of the atomic bomb from the perspective of their specialist fields. Atomic bomb survivor and author Ōta Yōko gave a special lecture at the festival. It reflected a growing desire on the part of Japanese to openly discuss the impact of the atomic bomb, in the latter stages of the Occupation. The exhibits and activities on the atomic bomb were sponsored by the Democratic Scientists Association and various departments of the university. Spread over four rooms, the exhibits provided information in the form of pictures and graphs on the effects of atomic energy on the human body, including the damage caused by the explosions at Hiroshima and Nagasaki. Student attendants gave brief talks on the history and development of atomic energy.[38] The atomic bomb exhibition attracted some 3000 people. Due to the positive response, the organizers decided to hold a wider-ranging exhibition in Kyoto city aimed at attracting members of the public.

Kyoto University thus came to sponsor a display of the paintings as part of a "Comprehensive Atomic Bomb Exhibition" ("Sōgō Genbaku-ten") at the Marubutsu Department Store near Kyoto station from 14–24 July 1951. It is testimony to the wide interest in the exhibition and the need for better understanding of the effects of the atomic bomb that Kyoto prefecture, Kyoto city and the Democratic Scientists Association were among the many sponsors. Entry was 30 yen and profits went to atomic bomb orphans.[39]

A-bombed items such as roof tiles and other historical materials documenting the destruction were borrowed from Hiroshima and Nagasaki. Local university students sought to convey the immensity of what occurred by producing a panorama of Hiroshima just after the bomb had been dropped. There were around 150 exhibition panels in the exhibition exploring aspects of the atomic bomb. For example, medical students produced poster panels discussing the recent academic literature on the medical effect of atomic bomb radiation on humans, and agriculture students summarized research on the genetic damage on plants exposed to radiation. Members of the Democratic Scientists Association contributed to the exhibition with explanations of the principles behind the atomic bomb and how they are made as well as touching on the peaceful uses of atomic energy.[40]

Close to the entrance to the exhibition were five of the Hiroshima panels that Akamatsu and Maruki had completed to date. The fourth and fifth panel had never been publicly exhibited before.[41] The public were particularly moved by the fifth panel showing boys and girls. The panels were augmented with drawings and *Pikadon* illustrations and the artists themselves who attended throughout the exhibition and made themselves available to sign various commemorative items. What is particularly significant is how the exhibition mobilized some 200 student volunteers and how the exhibition attracted around 30,000 visitors over ten days. By the end of the Kyoto exhibition, some 500,000 people had seen the Hiroshima panels over some 158 days.[42] But that was just the beginning. The Hiroshima panels continued to tour Japan.

The Kyoto exhibition encouraged students in other parts of Japan to do something similar. This can be seen when the Hiroshima panels travelled to Hokkaido. The Sapporo "Comprehensive Atomic Bomb Exhibition" ("Sōgō Genbaku-ten") was held as one part of the Hokkaido University Autumn Culture Festival. To encourage members of the public to attend, there were two outside venues: the fifth floor of the Sapporo Marui Imai Department Store and the long-established Fūkidō Bookstore. The main organizer was the university's cultural groups' federation with support from other students associations and the central committee. There was also widespread external support from more than thirty organizations including labour unions and women's groups.[43]

The Hiroshima panels and related drawings were shown firstly at the department store over two days, 20–21 November, attracting over 10,000 people. Due to their success, the artworks were moved to the university's central lecture theatre and shown for a further three days. Meanwhile, at the bookstore, there were panels and models produced by science students to explain the basics of nuclear physics and the principles underlying the production of atomic energy for peaceful purposes. The involvement of members of the Democratic Scientists Association ensured that there was also a plea for peace in the world. Engineering students contributed models and illustrations of nuclear reactors which helped to explain what power generation involved. Medical students explored the toll on human lives if the atomic bomb was dropped on Sapporo. Their fellow students in agriculture examined what the impact would be on domestic animals as well as how atomic energy might be usefully applied to benefit plants and animals. Other students contextualized the use of the atomic bomb within the history of war, how it contributed to an arms race, and others contemplated how the world would benefit if the funds being spent on nuclear weapons were spent instead on achieving peace. Students also explored how atomic energy would best be managed in future.[44]

The exhibition of the Hiroshima panels occurred against the backdrop of censorship of photographs and film footage of the atomic bomb as well as the oppression of the Japanese Communist Party that was adopting a new strategy of militancy and resorting to violence.[45] Paintings, however, seem to have avoided censorship even though the Hiroshima panels were bold in their depiction of the human cost of what occurred at Hiroshima and Nagasaki. Other Japanese artists had avoided depicting human beings but the Marukis drew attention to them, helping to make the destruction all too real.[46]

Civil Intelligence officials visited the exhibition in Sapporo. Unfortunately, three poster panels on display at the bookstore entitled "Shimin no koe" ("Voice of the Citizens") were deemed to contravene Occupation policy, and one of the organizers responsible for the display, Aoyagi Kiyoshi, was arrested and taken to Sapporo Police Headquarters. The panels had dared to suggest that the terrible use of the atomic bomb by the USA, a country so committed to democracy, made one question the nature of that democracy. There were also fears that the Marukis would also be arrested but this did not eventuate.[47] The matter did not end there.

On 21 January 1952, Sapporo Police Inspector Shiratori Kazuo was riding his bicycle home at around 7:30 p.m. and was shot dead by an assailant who then escaped the crime scene. Shiratori had been involved in counter measures to deal with the perceived threat posed by the Japanese Communist Party, including Aoyagi's arrest, so the JCP was suspected of involvement. As a result, Murakami Kuniji, a member of the JCP Sapporo Committee, was arrested and sentenced to twenty years' imprisonment.[48] Despite these problems, the curator Okamura Yukinori estimates that the Hiroshima panels were exhibited in some 170 exhibitions throughout Japan over 600 days between February 1950 and late 1953. The holding of the exhibitions and the public reception arguably constitute a social movement at the core of which was the visualization of the devastating force of the atomic bomb which had hitherto been concealed.[49] The Hiroshima panels made visible the suffering of the atomic bomb victims, providing an alternative narrative to public discourses about reconstruction and building a new Japan. However, the first ten panels would be sent overseas on an extended international tour from 1956 through to 1964. Unfortunately, this absence from Japan[50] coincided with the height of the promotion of the peaceful use of atomic energy in Japan.

Osada Arata and Children of the Atomic Bomb

While the Hiroshima panels attracted much attention, we can also point to other forms of visual production such as *kamishibai* (paper theatre or picture-story show) and films which provided opportunities for the Japanese to ponder the impact of atomic energy around this time. Osada Arata's edited collection entitled *Genbaku no ko: Hiroshima no shōnen to shōjo no uttae* (*Children of the Atomic Bomb: The Testament of the Boys and Girls of Hiroshima*) (1951) inspired both.

Osada experienced the atomic bomb himself. On 5 August 1945, Osada, a professor of education at Hiroshima Bunrika University (Hiroshima University of Science and Literature) stayed at the university on night duty. He had heard American reconnaissance planes flying over Hiroshima and the air-raid warnings that had issued forth. In the morning of 6 August, he wearily made his way back to his home in Hirano-cho, Hiroshima, just one mile from what would be the hypo-centre of the bomb that would be dropped soon after. He had a bath and sat on the porch, half-naked, drinking tea. This exposed him to the blast that morning and the shards of glass that resulted in fifty wounds to various parts of his body. His eighteen-year-old son, Gorō, fortunately happened to be visiting at that time, taking a break from his studies at Tokyo University of Commerce, what would later be known as Hitotsubashi University. Gorō was able to save his father who over the next four months would be close to death.[51] His father would go on to gather the recollections of children who experienced the atomic bomb. They were published in an edited collection entitled *Genbaku no ko: Hiroshima no shōnen to shōjo no uttae* (*Children of the Atomic Bomb: The Testament of the Boys and Girls of Hiroshima*) that was published by Iwanami Shoten in 1951 and translated into English in 1963.[52]

Osada's book is of special significance in that it inspired the production of an educational *kamishibai* entitled *Heiwa no chikai: Osada Arata hen "Genbaku no ko" yori* (*The Promise of Peace: From Osada Arata's "The Children of the A-Bomb"*). Consisting of sixteen panels, the script was written by the *kamishibai* pioneer Inaniwa Keiko[53] and illustrated by Satō Chūryō on his return from Siberia. The artist had been drafted into the armed forces and subsequently detained in Siberia from 1945 to 1948. He would go on to become one of Japan's most well-known sculptors. The *kamishibai* was published in 1952,[54] the final year of the Allied Occupation.

The *kamishibai* begins with the title panel followed by the dropping of the atomic bomb on Hiroshima on 6 August 1945 at 8:15 a.m. A piercing light flashes through the cloudless sky and a fire ball suddenly explodes. A second panel showing the devastated city of Hiroshima in flames is revealed to the audience. Clouds of dark smoke fill the sky and day becomes night. We are told that 247,000 people out of a total population of 400,000 died and children were forced into misery. The narrator explains that a few of the stories of these children will be told. The third panel depicts the story of Kenichi's mother who is pinned under some beams of her house. Kenichi and his father attempt to move the beams to no avail. With the fire spreading, the mother urges the father to save the children and escape. He tells Kenichi to run with his sister to somewhere safe. A fourth panel shows military officer and soldiers passing by but they ignore the father's desperate plea for help. The mother urges the father to escape the coming fire and join the children, but he refuses to leave his wife and prefers to die with her. The fifth panel shows a close-up of his forlorn face and the hand of his trapped wife from among the pillars. They are surrounded by smoke and flames. As her voice weakens, the father decides to leave to care for the children, his wife's dying wish.[55] The sixth panel shows him walking away, his hand covering his face in grief, against the backdrop of a devastated town on fire. The next panel reveals the image of the bent figure of a woman grasping her elbow with a boy by her side who is covering his ears trying to shut out the sounds of the conflagration. Both are in rags and seem like walking ghosts. In the background are other figures of people seeking to escape as well. We then go to the river where black, burnt corpses are floating on the surface.

The next panels tell the story of a young girl, Kiyoko who has a burn on her face. We see her first on a boat with her mother, bound for the first aid camp. A terribly injured girl offers them food. Six years later, Kiyoko is in eighth grade and writes down the story of the kind girl. She has a scar on her face and a few boys tease her. In the twelfth panel, Kenichi comes to Kiyoko's assistance and chastises the boys. Kiyoko then shows her scar and Kenichi urges them to take a good look at what the atomic bomb had given her. He points out how thousands of people had died in Hiroshima because of the bomb, dying even where they were standing. The boys then apologize and explain that they had come to Hiroshima after the war. Kiyoko and Kenichi think back to the events of 6 August. Kenichi points out that although the city of Hiroshima was being reconstructed, "the scars in our souls are not healed yet."[56] The second last panel shows a starry night in Hiroshima. In the foreground of the image is the A-Bomb Dome. Kenichi, his father and Kiyoko are probably sleeping. But the narrator tells us that in some dark corner there is the hum of an airplane and fears of another war. The children, we are told, are appealing to the world for peace. The last panel shows a hopeful image of doves of peace flying in the daylight above the up-held hands of children who have suffered from the atomic bomb. They wish fervently for peace and no more atomic bombs. In this way, through a succession of images, the emotive voice of the narrator and the stories from actual accounts, the *kamishibai* provided hand-drawn, moving images of what could otherwise not be described.

Osada's book would also be made into Shindō Kaneto's film *Genbaku no ko* (*Children of the Atomic Bomb*) (1952)[57] and Sekikawa (also known as Sekigawa) Hideo's *Hiroshima* (1953). In this way, Osada's text helped inspire some of the earliest examples of the visual representation of Hiroshima. The Japan Teachers Union was involved in commissioning both films. Indeed, the Japanese Communist Party would use *Heiwa no chikai* as a prime example of effective use of *kamishibai*. In the JCP's *Guide to Propaganda* (c. 1958) that was translated by the CIA and released in 2013,[58] JCP members would be instructed on the history and characteristics of the medium as well as how to best put on a show and to make the best use of *kamishibai*. This reinforced US fears that the anti-nuclear movement was linked to Communist and Communist-front organizations.

Conclusion

While censorship of the atomic bomb had a deleterious impact on the discussion and coming to terms of the Japanese people with what had happened at Hiroshima and Nagasaki, we can point to the role of Akamatsu Toshiko and Maruki Iri in the production and exhibition of the Hiroshima panels and collections like that compiled by Osada Arata in increasing awareness. With the end of the Occupation on 28 April 1952, it became possible to publish photographs of the atomic bombings without fear of censorship. The illustrated magazine *Asahi Gurafu* (*Asahi Graph*) duly did for its issue on 6 August 1952, on the occasion of the seventh anniversary of the dropping of the atomic bomb on Hiroshima. The special issue created a sensation among the public who had not been exposed to such graphic images of what had occurred. Twenty-two pages were devoted to showing the impact of the atomic bombs dropped on Hiroshima and

Nagasaki.[59] Five hundred and twenty thousand copies of the magazine quickly sold out.

But as important as all this was, it would not be until the Lucky Dragon No. 5 Incident in March 1954 when a tuna boat from Yaizu, Shizuoka prefecture, was exposed to radioactivity from a US hydrogen bomb test in the Pacific that a nationwide movement opposing the testing of atomic and hydrogen bombs would emerge. It was only then that all Japanese would feel threatened by the potential impact of nuclear weapons on their everyday life.

Notes

1. Yasuko Suga, "Modernism, Nationalism and Gender: Crafting 'Modern' Japonisme," *Journal of Design History* 21, no. 3 (Autumn 2008): 259–275.
2. Kamata Satoshi, "Gaka: Maruki Toshi (Gendai no shōzō)" ("Artist: Maruki Toshi, A Contemporary Portrait"), *Aera* (11 Aug. 1997): 52–56.
3. Kozawa Setsuko, *"Genbaku no zu": Egakareta "kioku," katarareta "kaiga"* (*"The Hiroshima Panels": Painted Memories, Narrated Paintings*) (Tokyo: Iwanami Shoten, 2002), 33.
4. Kozawa, *"Genbaku no zu,"* 35.
5. Usami Shō, *Ikebukuro monparunasu* (*Ikebukuro Montparnasse*) (Tokyo: Shūeisha, 1990), 21.
6. Toshima City Hall, "Nagasaki Atelier Village," accessed 15 Dec. 2019. https://www.city.toshima.lg.jp/artculture_en/brand/art/atelier-mura.html
7. Kozawa, *"Genbaku no zu,"* 36.
8. Kamata, "Gaka: Maruki Toshi (Gendai no shōzō);" Usami, *Ikebukuro monparunasu*, 456.
9. Maki Kaneko, "New Art Collectives in the Service of the War: The Formation of Art Organizations during the Asia-Pacific War," *positions: asia critique* 21, no. 2 (Spring 2013): 309–350.
10. Bert Winther-Tamaki, *Maximum Embodiment: Yoga, The Western Painting of Japan, 1912–1955* (Honolulu: University of Hawai'i Press, 2012), 8.
11. Charlotte Eubanks, "Avant-Garde in the South Seas: Akamatsu Toshiko's Micronesia Sketches," *Verge: Studies in Global Asias* 1, no. 2 (Fall 2015): 1–20, esp. 1, 17.
12. Tadao Yanaihara, *Pacific Islands under Japanese Mandate* (London: Oxford University Press, 1940), 280–305.
13. Mark R. Peattie, *Nan'yō: The Rise and Fall of the Japanese in Micronesia, 1885–1945* (Honolulu: University of Hawai'i Press, 1988).

14. Okaya Kōji, *Nankai hyōtō: Mikuroneshia ni miserareta Hijikata Hisakatsu, Sugiura Sasuke, Nakajima Atsushi* (*South Seas Drifting*: Hijikata Hisakatsu, Sugiura Sasuke and Nakajima Atsushi) (Tokyo: Fuzanbō Intānashonaru, 2007), 132–133.
15. Kozawa, *"Genbaku no zu,"* 38.
16. "Inochi o mitsumete: Maruki Toshi, seitan 100-nen, chū, Nan'yō taiken to shūsen" ("Gazing at Life: 100th Anniversary of the Birth of Maruki Toshi, part 2, Her South Seas Experience and the End of the War"), *Chūgoku shimbun* (*Chūgoku Newspaper*), morning edition, 2 August 2012, http://www.hiroshimapeacemedia.jp/?p=22185
17. Nishi Haruhiko, *Kaisō no Nihon gaikō* (*Reminiscences of Japanese Diplomacy*) (Tokyo: Iwanami Shoten, 1965), 102–111.
18. Kozawa, *"Genbaku no zu,"* 46.
19. Ida Yoshiharu and Nishi Teruhiko, eds, "Nishi Haruhiko ryaku nenpu chosho ronbun nado mokuroku shō" ("Short Chronology and Selected Publications of Nishi Haruhiko," *Eigaku-shi kenkyū* (*History of English Studies*) 20 (1987): 241–245; Kozawa, *"Genbaku no zu,"* 46.
20. Kamata, "Gaka: Maruki Toshi (Gendai no shōzō)."
21. Takaaki Kumagai, "Kitagawa Tamiji's Art and Art Education: Translating Culture in Postrevolutionary Mexico and Modern Japan," PhD dissertation, University of Kansas, 2017, 141.
22. Usami, *Ikebukuro monparunasu*, 452.
23. Maruyama Kaoru, illustrated by Akamatsu Toshiko, *Yashi no mino tabi* (*The Travels of a Palm Tree Fruit*) (Tokyo: Teikoku Kyōiku Shuppanbu, 1942).
24. Kamata, "Gaka: Maruki Toshi (Gendai no shōzō);" Kozawa, *"Genbaku no zu,"* 53.
25. Tsuchiya Yukio, illustrated by Akamatsu Toshiko, *Yashi no ki no shita* (*Under a Palm Tree*) (Tokyo: Shōgakkan, 1942).
26. Kamata, "Gaka: Maruki Toshi (Gendai no shōzō)."
27. Tsuchiya, *Yashi no ki no shita* (*Under a Palm Tree*).
28. Kozawa Setsuko, "Maruki Suma to Daidō Aya no `kaiga sekai'" (The Artistic World of Maruki Suma and Daidō Aya), *Genbaku bungaku kenkyū* (*Atomic Bomb Literature Studies*) 8 (21 Dec. 2009): 169–182.
29. Kamata, "Gaka: Maruki Toshi (Gendai no shōzō);" Richard H. Minear, "The Hiroshima Murals of Maruki Iri and Maruki Toshi: A Note," in *Hiroshima: Three Witnesses*, ed. and trans. Richard H. Minear (Princeton: Princeton University Press, 1990), 371–378, esp. 374.
30. Cleo Macmillan, "The Power of the Hiroshima Panels," *Arena Magazine* no. 121 (Dec. 2012 - Jan 2013): 29–34; "The Hiroshima Panels," *Australian Left Review* 1, no. 2 (Aug. – Sept. 1966): 32–33.

31. Both were ousted from the Japanese Communist Party in 1964 over their opposition to all nuclear tests, including those of the Soviet Union. Yukinori Okamura, "The Hiroshima Panels Visualize Violence: Imagination over Life," *Journal for Peace and Nuclear Disarmament* 2, no. 2 (2019): 518–534, esp. 520, 528.
32. Okamura, "The Hiroshima Panels Visualize Violence," note 3.
33. Okamura, "The Hiroshima Panels Visualize Violence," esp. 522.
34. GHQ, UN and Far East Command, Military Intelligence Section, General Staff, *Intelligence Summary: Intelligence Data covering the Military Counterintelligence, Economic and Political Fields in: Japan, Korea, Philippines, China, Southeast Asia*, Secret, no. 3252 (5 Aug. 1951), Japan Section, J-2. It has been suggested that the Hiroshima panels constitute one component of a performance-based art form. See Charlotte Eubanks, "The Mirror of Memory: Constructions of Hell in the Marukis' Nuclear Murals," *PMLA* (Modern Language Association of America) 124, no. 5 (Oct. 2009): 1614–1631, esp. 1615.
35. GHQ, UN and Far East Command, Military Intelligence Section, General Staff, *Intelligence Summary: Intelligence Data covering the Military Counterintelligence, Economic and Political Fields in: Japan, Korea, Philippines, China, Southeast Asia*, Secret, no. 3253 (6 Aug. 1951), Japan Section, J-2.
36. It was subsequently reorganized and became the Japan Peace Protection Committee (Heiwa Yōgo Nippon Iinkai). See Nagashima Yūki, "Heiwa yōgo undo ni okeru tōronshūkai no keisei (1952–1953 nen") ("The Formation of the Peace Protection Movement Forum, 1952–1953"), *Ōhara Shakai Mondai Kenkyūjo zasshi* (*Journal of Ōhara Institute for Social Research*), no. 709 (Nov. 2017): 44–57, esp. 46.
37. GHQ, UN and Far East Command, Military Intelligence Section, General Staff, *Intelligence Summary: Intelligence* (6 Aug. 1951), Japan Section, J-2.
38. GHQ, UN and Far East Command, Military Intelligence Section, General Staff, *Intelligence Summary: Intelligence Data covering the Military Counterintelligence, Economic and Political Fields in: Japan, Korea, Philippines, China, Southeast Asia*, Secret, no. 3254 (7 Aug. 1951), Japan Section, J-2.
39. Okamura Yukinori, *"Genbaku no zu" zenkoku junkai: Senryōka, hyakuman-nin ga mita!* (*The Nationwide Tour of the Hiroshima Panels: One Million People Saw Them!*) (Tokyo: Shinjuku Shobō, 2015), 104–105.
40. Okamura, *"Genbaku no zu" zenkoku junkai*, 104.
41. Kawai Ichirō, "Senryōka no `Sōgō genbaku-ten' 3" ("Occupation-Period `Comprehensive Atomic Bomb Exhibition', Part 3," *Kyōto hoken i shimbun* (*Kyoto Medical Practitioners Newspaper*), 20 July, 2011, 4.
42. Okamura, *"Genbaku no zu" zenkoku junkai*, 110.

43. Okamura, *"Genbaku no zu" zenkoku junkai*, 119.
44. Okamura, *"Genbaku no zu" zenkoku junkai*, 120–121.
45. Paul F. Langer, *Communism in Japan: A Case of Political Naturalization* (Stanford: Hoover Institution Press, 1972), 49.
46. Okamura, "The Hiroshima Panels Visualize Violence," esp. 521.
47. Okamura, *"Genbaku no zu" zenkoku junkai* 123–125.
48. Okamura, *"Genbaku no zu" zenkoku junkai,* 125. For further details of the case, see Imanishi Hajime, "Shinpojiumu: Rekishi toshite no Shiratori jiken, Shiratori jiken to wa nani ka" ("Symposium: The Shiratori Incident as History, What was the Shiratori incident?"), *Shōgaku tōkyū* (*Otaru University of Commerce Business Studies*) 64, nos. 2–3 (2012): 3–19.
49. Okamura, "The Hiroshima Panels Visualize Violence," esp. 523.
50. Inaga Shigemi, "Sensō-ga to heiwa-ga no aida: Rekishi no naka no kaiga sakuhin no unmei: Maruki Iri, Toshi fusai "Genbaku no zu" saikō, jyō" ("Somewhere between War Paintings and Peace Paintings: The Fate of Paintings in History, A Reconsideration of Maruki Iri and Toshi's Hiroshima Panels, Part 1"), *Aida*, no. 113 (20 May 2005): 39–44.
51. Gorō Osada, "Messages from Hiroshima," Asahi Shimbun Digital, *Memories of Hiroshima and Nagasaki: Messages from Hibakusha*, accessed 3 Jan. 2020. http://www.asahi.com/hibakusha/english/hiroshima/h00-00009e.html
52. Arata Osada, comp., *Children of the A-Bomb: The Testament of the Boys and Girls of Hiroshima*, trans. by Jean Dan and Ruth Sieben-Morgen (London: Peter Owen, 1963).
53. Kusakabe Shigeko, "Kamishibai undo o kiri hiraita hitobito: Inaniwa Keiko, heiwa o negatte" ("Pioneers of the Kamishibai Movement: Inaniwa Keiko, Wishing for Peace," *Rikkyō Daigaku Jogakuin Tanki Daigaku kiyō* (*Rikkyō University Women's College Bulletin*), no. 47 (2015): 109–122, esp. 116–117.
54. Inaniwa Keiko, illustrated by Satō Chūryō, *Heiwa no chikai: Osada Arata hen "Genbaku no ko" yori* (*The Promise of Peace: From Osada Arata's "Children of the A-Bomb"*) (Tokyo: Kyōiku Kamishibai Kenkyūkai, 1952).
55. See *Kamishibai, "Heiwa no chikai"* (*Paper Theatre "The Promise of Peace"*), accessed 2 Jan. 2020 https://www.youtube.com/watch?v=wov7rtdK3xs. For a written description in English, Japan Communist Party, comp., *Guide to Propaganda* (c. 1958), 117–127, General CIA Records, CIA-RDP81-01043R002300230001-6.
56. See *Kamishibai, "Heiwa no chikai"*; Japan Communist Party, comp., *Guide to Propaganda.*
57. Yuko Shibata, *Producing Hiroshima and Nagasaki: Literature, Film, and Transnational Politics* (Honolulu: University of Hawai'i Press, 2018), 46.
58. Japan Communist Party, comp., *Guide to Propaganda.*

59. Hirofumi Utsumi, "Nuclear Images and National Self-Portraits: Japanese Illustrated Magazine *Asahi Graph*, 1945–1965," *Annual Review of the Institute for Advanced Social Research* 5 (March 2011): 1–29, esp. 8–9.

Bibliography

Eubanks, Charlotte. "Avant-Garde in the South Seas: Akamatsu Toshiko's Micronesia Sketches." *Verge: Studies in Global Asias* 1, no. 2 (Fall 2015): 1–20, esp. 1, 17.

Eubanks, Charlotte. "The Mirror of Memory: Constructions of Hell in the Marukis' Nuclear Murals." *PMLA* (Modern Language Association of America) 124, no. 5 (Oct. 2009): 1614–1631.

GHQ, UN and Far East Command, Military Intelligence Section, General Staff. *Intelligence Summary: Intelligence Data covering the Military Counterintelligence, Economic and Political Fields in: Japan, Korea, Philippines, China, Southeast Asia*. Secret, no. 3252 (5 Aug. 1951), Japan Section, J-2; no. 3253 (6 Aug. 1951), Japan Section, J-2.

Ida Yoshiharu and Nishi Teruhiko, eds. "Nishi Haruhiko ryaku nenpu chosho ronbun nado mokuroku shō" ("Short Chronology and Selected Publications of Nishi Haruhiko." *Eigaku-shi kenkyū* (*History of English Studies*) 20 (1987): 241–245.

Imanishi, Hajime. "Shinpojiumu: Rekishi toshite no Shiratori jiken, Shiratori jiken to wa nani ka" ("Symposium: The Shiratori Incident as History, What was the Shiratori incident?"). *Shōgaku tōkyū* (*Otaru University of Commerce Business Studies*) 64, nos. 2–3 (2012): 3–19.

Inaga, Shigemi. "Sensō-ga to heiwa-ga no aida: Rekishi no naka no kaiga sakuhin no unmei: Maruki Iri, Toshi fusai "Genbaku no zu" saikō, jyō" ("Somewhere between War Paintings and Peace Paintings: The Fate of Paintings in History, A Reconsideration of Maruki Iri and Toshi's Hiroshima Panels, Part 1"). *Aida*, no. 113 (20 May 2005): 39–44.

Inaniwa, Keiko, illustrated by Satō Chūryō. *Heiwa no chikai: Osada Arata hen "Genbaku no ko" yori* (*The Promise of Peace: From Osada Arata's "Children of the A-Bomb"*). Tokyo: Kyōiku Kamishibai Kenkyūkai, 1952.

"Inochi o mitsumete: Maruki Toshi, seitan 100-nen, chū, Nan'yō taiken to shūsen" ("Gazing at Life: 100th Anniversary of the Birth of Maruki Toshi, part 2, Her South Seas Experience and the End of the War"). *Chūgoku shimbun* (*Chūgoku Newspaper*), morning edition, 2 August 2012, http://www.hiroshimapeacemedia.jp/?p=22185.

Japan Communist Party, comp. *Guide to Propaganda* (c. 1958), 117–127, General CIA Records, CIA-RDP81-01043R002300230001-6.

Kamata, Satoshi. "Gaka: Maruki Toshi (Gendai no shōzō)" ("Artist: Maruki Toshi, A Contemporary Portrait"). *Aera* (11 Aug. 1997): 52–56.

Kamishibai, "Heiwa no chikai" (*Paper Theatre "The Promise of Peace"*). Accessed 2 Jan. 2020. https://www.youtube.com/watch?v=wov7rtdK3xs.

Kawai, Ichirō. "Senryōka no `Sōgō genbaku-ten' 3" ("Occupation-Period `Comprehensive Atomic Bomb Exhibition', Part 3." *Kyōto hoken i shimbun* (*Kyoto Medical Practitioners Newspaper*), 20 July, 2011.

Kozawa, Setsuko. *"Genbaku no zu": Egakareta "kioku," katarareta "kaiga"* (*"The Hiroshima Panels": Painted Memories, Narrated Paintings*). Tokyo: Iwanami Shoten, 2002.

Kumagai, Takaaki. "Kitagawa Tamiji's Art and Art Education: Translating Culture in Postrevolutionary Mexico and Modern Japan." PhD dissertation, University of Kansas, 2017.

Kozawa, Setsuko. "Maruki Suma to Daidō Aya no `kaiga sekai'" (The Artistic World of Maruki Suma and Daidō Aya). *Genbaku bungaku kenkyū* (*Atomic Bomb Literature Studies*) 8 (21 Dec. 2009): 169–182.

Kusakabe, Shigeko. "Kamishibai undo o kiri hiraita hitobito: Inaniwa Keiko, heiwa o negatte" ("Pioneers of the Kamishibai Movement: Inaniwa Keiko, Wishing for Peace"). *Rikkyō Daigaku Jogakuin Tanki Daigaku kiyō* (*Rikkyō University Women's College Bulletin*), no. 47 (2015): 109–122.

Langer, Paul F. *Communism in Japan: A Case of Political Naturalization*. Stanford: Hoover Institution Press, 1972.

Macmillan, Cleo. "The Power of the Hiroshima Panels." *Arena Magazine* no. 121 (Dec. 2012 – Jan 2013): 29–34.

Maki, Kaneko. "New Art Collectives in the Service of the War: The Formation of Art Organizations during the Asia-Pacific War." *positions: asia critique* 21, no. 2 (Spring 2013): 309–350.

Maruyama, Kaoru, illustrated by Akamatsu Toshiko. *Yashi no mino tabi* (*The Travels of a Palm Tree Fruit*). Tokyo: Teikoku Kyōiku Shuppanbu, 1942.

Minear, Richard H. "The Hiroshima Murals of Maruki Iri and Maruki Toshi: A Note." In *Hiroshima: Three Witnesses*, edited and trans. by Richard H. Minear, 371–378. Princeton: Princeton University Press, 1990.

Nagashima, Yūki. "Heiwa yōgo undo ni okeru tōronshūkai no keisei (1952–1953 nen") ("The Formation of the Peace Protection Movement Forum, 1952–1953"). *Ōhara Shakai Mondai Kenkyūjo zasshi* (*Journal of Ōhara Institute for Social Research*), no. 709 (Nov. 2017): 44–57.

Nishi, Haruhiko. *Kaisō no Nihon gaikō* (*Reminiscences of Japanese Diplomacy*). Tokyo: Iwanami Shoten, 1965.

Okamura, Yukinori. *"Genbaku no zu" zenkoku junkai: Senryōka, hyakumannin ga mita!* (*The Nationwide Tour of the Hiroshima Panels: One Million People Saw Them!*). Tokyo: Shinjuku Shobō, 2015.

Okamura, Yukinori. "The Hiroshima Panels Visualize Violence: Imagination over Life." *Journal for Peace and Nuclear Disarmament* 2, no. 2 (2019): 518–534.

Okaya, Kōji. *Nankai hyōtō: Mikuroneshia ni miserareta Hijikata Hisakatsu, Sugiura Sasuke, Nakajima Atsushi* (*South Seas Drifting*: Hijikata Hisakatsu, Sugiura Sasuke and Nakajima Atsushi). Tokyo: Fuzanbō Intānashonaru, 2007.
Osada, Arata, comp. *Children of the A-Bomb: The Testament of the Boys and Girls of Hiroshima*, trans. by Jean Dan and Ruth Sieben-Morgen. London: Peter Owen, 1963.
Osada, Gorō. "Messages from Hiroshima." Asahi Shimbun Digital, *Memories of Hiroshima and Nagasaki: Messages from Hibakusha*. Accessed 3 Jan. 2020. http://www.asahi.com/hibakusha/english/hiroshima/h00-00009e.html.
Peattie, Mark R. *Nan'yō: The Rise and Fall of the Japanese in Micronesia, 1885–1945*. Honolulu: University of Hawai`i Press, 1988.
Shibata, Yuko. *Producing Hiroshima and Nagasaki: Literature, Film, and Transnational Politics*. Honolulu: University of Hawai`i Press, 2018.
Suga, Yasuko. "Modernism, Nationalism and Gender: Crafting `Modern' Japonisme." *Journal of Design History* 21, no. 3 (Autumn 2008): 259–275.
"The Hiroshima Panels." *Australian Left Review* 1, no. 2 (Aug. – Sept. 1966): 32–33.
Toshima City Hall, "Nagasaki Atelier Village." Accessed 15 Dec. 2019. https://www.city.toshima.lg.jp/artculture_en/brand/art/atelier-mura.html.
Tsuchiya, Yukio, illustrated by Akamatsu Toshiko. *Yashi no ki no shita* (*Under a Palm Tree*). Tokyo: Shōgakkan, 1942.
Usami, Shō. *Ikebukuro monparunasu* (*Ikebukuro Montparnasse*). Tokyo: Shūeisha, 1990.
Utsumi, Hirofumi. "Nuclear Images and National Self-Portraits: Japanese Illustrated Magazine *Asahi Graph*, 1945–1965." *Annual Review of the Institute for Advanced Social Research* 5 (March 2011): 1–29.
Winther-Tamaki, Bert. *Maximum Embodiment: Yoga, The Western Painting of Japan, 1912–1955*. Honolulu: University of Hawai`i Press, 2012.
Yanaihara, Tadao. *Pacific Islands under Japanese Mandate*. London: Oxford University Press, 1940.

CHAPTER 4

The Beginnings of Atoms for Peace in Japan

Introduction

We saw in Chap. 2 how US pre-surrender planning had leaned towards retaining the Japanese Emperor, how a show of force was seen by American leaders as being needed to end the war, and how Emperor Hirohito's surrender speech suggested that the nation's defeat had been a result of superior American science and technology. In the aftermath of the atomic bombs that were dropped on Hiroshima and Nagasaki, Japanese scientists and film-makers scrambled to record the damage that had been caused by the nuclear weapons. Censorship during the subsequent Allied Occupation of Japan (1945–1952), which was led by Americans, meant that images of the destruction and critique of the bomb were severely hampered. As we saw in Chap. 3, this didn't prevent Akamatsu Toshiko and Maruki Iri from painting the Hiroshima panels but the documentary and film footage that scientists and film-makers had taken was removed to the USA and not made available to the Japanese until the late 1960s.

This was in the interests of maintaining harmonious relations between the USA and Japan and ensuring that Japan would remain as a bulwark against the spread of Communism in Asia. The A-Bomb Dome had been the Hiroshima Prefectural Industrial Promotion Hall and in a way, this resonated with post-war Japan's more pacifist stance. By embracing science, technology and trade, especially with the USA, Japan would face the future. There were efforts to do this by scientific exchange by inviting

M. Low, *Visualizing Nuclear Power in Japan*, Palgrave Studies in the History of Science and Technology,
https://doi.org/10.1007/978-3-030-47198-9_4

eminent physicists such as Yukawa Hideki to the USA and by promoting trade between the two nations. We can point to the America Fair in Nishinomiya city, 18 March to 31 May 1950, and the Kobe Fair, 15 March to 15 June 1950 as early examples. This was discussed in Chap. 2. It was not until the Lucky Dragon Incident of 1954 when tuna fishermen were exposed to radioactive fallout from an American hydrogen bomb test at Bikini Atoll that the Japanese public experienced widespread nuclear fear that inspired the film *Godzilla* (1954). This will be discussed in Chaps. 5 and 6. There was a perception that Japan had become the victims of yet another nuclear weapon and this gave rise to considerably anti-American feeling.

The promotion of "Atoms for Peace" in Japan sought to offset this, to enhance science and technology ties, promote trade and enhance the political career prospects of the owner of the newspaper *Yomiuri Shimbun*, Shōriki Matsutarō who sought to use his newspaper and influence to elevate himself all the way up to the position of Prime Minister. This chapter outlines his efforts to bring atomic energy for power generation to Japan. No single person influenced Japanese nuclear culture more than Shōriki. His vision of nuclear power in Japan owes much to the USA and to Great Britain.

Towards an Atomic Marshall Plan

On 2 February 1950, Senator Brien McMahon (Democrat, Connecticut), author of the 1946 Atomic Energy Act, first chairman of the Joint Committee on Atomic Energy and a key figure in the establishment of the Atomic Energy Commission, spoke in the Senate, calling for a $50 billion global Marshall Plan that would include the development of atomic energy for peaceful purposes and economic aid. He suggested that

> Perhaps through atomic power for industry and agriculture we can transform the deserts of Africa, Asia, and the Americas into blooming crop-producing acres, and the arid hills of the world into gardens. It is almost impossible to overestimate what all-out concentration upon atomic energy for peace might accomplish in terms of remaking and improving the physical environment of mankind.[1]

In Japan, there was almost a sense of entitlement about how Japan deserved to exploit the atom given what they had experienced at Hiroshima and Nagasaki. In 1952, the physicist Taketani Mituo (pronounced Mitsuo) wrote an article entitled "The Direction of Atomic Energy Research" which appeared in the November issue of the magazine *Kaizō*. He argued that

> The Japanese, being the only casualties of atomic warfare, are entitled to have the strongest say in the development of atomic energy. We, the Japanese, must carry out atomic energy research ourselves, but we must never aspire to weapons research, in order to appease the spirits of the atomic bomb victims.[2]

According to Taketani, Japan had the greatest moral right to conduct research into the peaceful use of atomic energy and other nations should assist Japan's efforts, including by supplying uranium.

On 22 October 1953, US Atomic Energy Commissioner Thomas E. Murray announced before a convention of electric company officials in Chicago, that the Commission would construct the nation's first full-scale nuclear power plant in what would be "America's peacetime answer to recent Soviet atomic weapon tests."[3] He feared that "the USSR had learned how to make atomic industrial power, and was ready to supply the know-how to other countries in exchange for uranium and other economic concessions." "I believe that unless we make all-out attack on our nuclear power program immediately, we may be deprived of foreign uranium ores, with the result that our weapons potential will be smaller than need be the case."[4] In this way, he connected the nuclear arms race with the race to realize nuclear industrial power. Although government would carry the financial burden initially, as technology improved, this could be increasingly transferred to private industry.[5]

Less than two months later, President Dwight D. Eisenhower delivered a speech before the United Nations General Assembly in New York City on 8 December 1953. In his speech, he announced a new policy that the USA would implement to promote the peaceful applications of nuclear technology. He proclaimed that "It is not enough to take this weapon out of the hands of the soldiers. It must be put into the hands of those who will know how to strip its military casing and adapt it to the arts of peace."[6] He pledged the USA to devoting "its entire heart and mind" to finding the way by which "the miraculous inventiveness of man shall not be dedicated to his death, but consecrated to his life."[7] The US Mission to the UN reported that many delegates responded very positively to Eisenhower's speech. It was seen by many as the most significant and important address given by the USA at the UN.[8] A small number expressed scepticism, suggesting that it would not amount to much given that it would not bring about disarmament. Overall though, it seems to have had the desired effect of communicating that the USA was not a war-monger, and the UN Secretary General Dag Hammarskjöld felt that it had received widespread approval.[9]

Eisenhower's speech came as a surprise to members of the US Atomic Energy Commission including Murray whose speech it resonated with. Chairman Lewis L. Strauss had been consulted but no other commissioners had seen the proposal.[10] Henry D. Smyth, a nuclear physicist who had participated in the Manhattan Project and was a member of the US Atomic Energy Commission, saw Atoms for Peace as a "thoroughly dishonest proposal" given the danger of nuclear proliferation through the production of plutonium, and because it exaggerated the prospects at the time for nuclear power.[11]

In January 1954, at the 38th Meeting of the General Advisory Committee to the US Atomic Energy Commission, the physicist and Nobel laureate I.I. Rabi remarked that underlying Eisenhower's proposals was the premise "that nuclear power was practical and desirable in the almost immediate future," adding that "we would have to be able to live up to this."[12] Fellow physicist Eugene P. Wigner thought that Eisenhower's Atoms for Peace speech was wonderful but was concerned that implementation would lead to a second Marshall plan. Rabi and Wigner both urged the USA to speed up their reactor program.[13]

By early August 1954, the proposed US Atoms for Peace program had largely become a scheme to assist selected countries construct and use atomic research reactors through bilateral agreements. The cost of construction and operation would be borne by the participating countries and it was proposed that reactor operation would be carried out under the authority of a proposed UN International Atomic Energy Agency.[14] The CIA Office of National Estimates anticipated that the proposed program would be criticized by some as being too modest and that there would be calls for it to assist in the installation of power reactors rather than research reactors. The Office suggested that it should be emphasized that the provision of research reactors were "an essential prerequisite to wider peacetime uses of atomic energy and that assistance in the development of power reactors could be expected to follow in due course."[15] Peaceful uses of atomic energy were still at the experimental stage anyway, so a largely research reactor-focused program would provide an opportunity to train the requisite scientific and technical personnel who would be needed later down the line.

The CIA Office of National Estimates foresaw, in 1954, that some countries might become disillusioned as they became aware of the cost of maintaining an atomic energy research program, leading to calls for further US financial assistance. There might also be popular dissatisfaction with the "lack of tangible and spectacular benefits," so care needed to be taken against "excessive expectations."[16] Given the President's rhetoric about how the Atoms for Peace was for the benefit of mankind, the Office anticipated difficulties in tying reciprocal, political commitments to the USA through bilateral assistance. The eventual establishment of an International Atomic Energy Agency might ease the way for prior bilateral agreements. Japan was seen as being one of a group of Free World countries that also included Italy, India, Argentina, Brazil, Australia and South Africa that were less scientifically advanced than countries like the United Kingdom, France and West Germany, and would require considerable aid and assistance if they were to make effective use of a research reactor. The Office also anticipated that the Soviet Union would likely point out that the USA was nevertheless planning to continue full-scale production of atomic weapons and designed to "secure US control over atomic research and production throughout the world."[17]

Commissioner Murray gave a major policy address before the seventh biennial convention of the United Steelmakers of America on 21 September 1954. He argued that the USA should build a nuclear power reactor in Japan lest the Soviet Union gain the upper hand and force countries to pay a high price, not only in terms of the reactor itself but in other concessions as well. He called for the government to take action "because the economics of nuclear power are so uncertain it is unrealistic to expect private industry to undertake, on a purely risk basis, anything like the effort that the world atomic power problem demands."[18] He proposed Japan because it was "the first and is still the only land which has been engulfed in the white flame of the atom."[19] He unknowingly reinforced the position that the physicist Taketani and other Japanese were taking in regard to nuclear power in Japan. Indeed, Murray felt that they should act "while the memory of Hiroshima and Nagasaki remains so vivid." An American power reactor would be a "lasting monument to our technology and our goodwill." Not only this, but Murray felt that such a gesture would not only be "a dramatic and Christian gesture" but that it would go some way to nullifying what had just been lost in terms of prestige at the first International Conference on the Peaceful Uses of Atomic Energy in August 1955 at Geneva, Switzerland, where the Soviets and British advances in power reactors had arguably outshone American achievements.[20]

Murray explained to the United Steelworkers of America convention that there were essentially two types of reactors: those primarily tasked with producing power and the other aimed at producing plutonium with electric power as a by-product. The latter, dual-purpose reactor might seem attractive to private industry, but Murray felt that the US government already had sufficient facilities for weapons-grade plutonium. What's more, it is unlikely that the USA would want to place plutonium plus power reactors in foreign countries.[21] He spoke of the atomic dilemma, namely how the atom represented world destruction on the one hand and peaceful power for mankind on the other. The two were not unconnected. Indeed, he assured the Steelworkers that "we must do the one without neglecting the other."[22]

We can discern from Murray's speech his desire to ensure that the USA remained a leader in nuclear technology, his anti-Communist stance, his clear thinking as an engineer, and his strong religious beliefs as a practicing Catholic.[23] Building a power reactor in Japan made good sense from all these viewpoints. It also made good sense in terms of diplomacy. The *Washington Post* praised Murray's idea, especially in view of the pending arrival in November of Prime Minister Yoshida Shigeru. There was a perception that the Soviets would view the urgent energy needs of the Japanese as opportunities to make political overtures. The *Washington Post* agreed that Japan deserved to be the first foreign recipient of power reactor given that Americans were now aware, thanks to the US bombing survey, that "the dropping of the atomic bombs on Japan was not necessary." The *Post* argued "how better to make a contribution to amends than by offering Japan the means for the peaceful utilization of atomic energy." At the same time, it would serve to divert attention from the arms race.[24] Indeed, when Yoshida met with President Eisenhower in Washington, DC, Eisenhower "expressed the hope that progress would be made in developing the peaceful uses of atomic energy" but that it was unfortunate that the lack of Soviet cooperation meant that "a major share of atomic capabilities" were devoted to destructive purposes.[25]

The Christian Science Monitor interpreted Murray's proposal as calling for the reactor to be built in Hiroshima. The atom would thereby return to Hiroshima in a constructive role that would encourage the USA to focus more on nuclear power than it had hitherto done, and to communicate to the Japanese and the peoples of the world that there are Americans who are more enthusiastic about the peaceful uses of atomic energy than in how large a bomb can be made.[26]

On 1 December 1954, John Jay Hopkins, President, General Dynamics Corporation, New York, proposed an "Atomic Marshall Plan."[27] General Dynamics had just built the world's first operational nuclear-powered submarine, the USS Nautilus so Hopkins was seen as having strong credentials. He began by expressing dismay that the Soviets had surpassed the USA in the industrial application of atomic energy. He feared that the USA would be doomed to an inferior competitive position, falling behind even Communist China due to its unlimited access to cheap labour. He felt that Asia could not be abandoned to control by the Soviet Union or Communist China. An Atomic Marshall plan was called for. He proposed that a hundred-year program for the financing and construction of reactors be established which would involve American private enterprise working with the US government and the governments of friendly countries. It would begin with "atomic seed reactors" and be funded by diverting some of the pre-existing economic aid to the provision of reactors.

Hopkins pointed out how research reactors were relatively inexpensive, ranging in price from several 100,000 dollars up to 3 million dollars. In addition, they could produce valuable radioisotope by-products that could be used in industry, medicine and agriculture. Research reactors would thus serve as the first step in the introduction of industrial atomic reactors. This might be followed by the provision of portable atomic power reactors. But the main objective would be large, stationary atomic power plants such as the 60,000 kilowatt pressurized water reactor being developed by Westinghouse. Power plants were expensive and required a workforce of highly trained technicians and engineers. Hopkins admitted that it was still early days in terms of optimal design and the provision of economical electric power.

The ultimate step would be the construction of large-scale breeder reactors that effectively bred their own replacement fuel and produced fuel for other reactors. Hopkins mentioned how Japan had been given $2.276 billion in non-military economic aid and credits between 1945 and 1952. If this had been spent on the installation of American industrial atomic reactors, Japan, Hopkins argued, would have around hundred large atomic power plants with the capacity to produce 10 million kilowatts of power. This would almost match the total installed capacity of 12 million kilowatts in 1952.[28]

American industry would profit from an Atomic Marshall Plan and the relationships that were created would create other opportunities and outlets for American enterprises and products. Atomic energy could provide "the last opportunity we possess to create a stable world."[29] Indeed, Hopkins was hoping for an economic chain reaction with reactors helping to establish industry in underdeveloped nations, creating jobs, and producing surplus capital that could be reinvested and result in rising living standards. If the USA didn't help provide reactors, nations could turn to Europe or the Soviet Union. We thus see by the end of 1954, support in the USA coalescing around the idea of providing a reactor to Japan. In January 1955, *The Los Angeles Times* wrote glowingly "On Spreading the Atom Around," claiming that "our reactors would be blessed monuments, commending our ways, just as roads and aqueducts gave prestige to the ways of the Romans."[30] Whereas the proposed Atoms for Peace program was leaning towards small, research reactors, politicians and public commentators were hoping for a larger gesture in the form of assistance towards the construction of a power reactor. The perception was that this would be beyond private firms to provide and that the government would need to step in. There was some agreement that Japan would be a worthy and appropriate recipient and the idea spread that Hiroshima would be highly symbolic location.

Meanwhile, in Japan, news of Hopkins' proposal had reached media mogul and politician Shōriki Matsutarō, owner of the newspaper *Yomiuri Shimbun*. Shōriki communicated to the USA that he was keen to engage in a fully fledged Atoms for Peace campaign in Japan and was seeking American guidance and support. He reported that Hopkins' speech had considerable impact among Japanese industrialists, most notably his comments urging installation of atomic power plants in Japan and East Asia in what had been dubbed an "Atomic Marshall Plan." The Japanese were struck by the practical tone of Hopkins' approach. William Halsted, President of Unitel Corp., had been in touch with Shōriki on behalf of Hopkins, in regard to the possibility of Hopkins visiting Japan to explain his proposal. This was part of an attempt by Shōriki to invite "atomic power peace emissaries" to Japan where Shōriki promised to provide publicity and to introduce them to leading Japanese government and industry representatives.[31]

Atomic Energy for Everyone

On 6 August 1954, the city of Hiroshima held an official peace ceremony to commemorate the anniversary of the dropping of the first atomic bomb on a mostly civilian population. At 8:15 a.m. when the bomb exploded over Hiroshima nine years before, sirens sounded and church bells rang throughout the city. Five hundred white doves were released[32] and a peace rally involving labour unions and women's organizations opposing atomic bombs and hydrogen bombs followed. The physicist Yukawa Hideki conveyed a message to the rally supporting a ban on nuclear weapons. He emphasized that it was vital to ensure that atomic energy was used for the benefit and advancement of human welfare.[33] As the editorial of the *Nippon Times* put it, "today's anniversary is one of poignant memories although not one unmixed with hope for the future."[34] Day Inoshita, writing for the same newspaper, noted that "there was surprisingly little bitterness among the majority of the Hiroshima residents toward the United States."[35] In this way, the push for the development of nuclear power was adroitly made compatible with what had occurred at Hiroshima, Nagasaki and at Bikini Atoll. It did not negate the terrible destruction which occurred but the narrative was that the Japanese could now look to the future and that the USA could assist them in their quest to harness atoms for peaceful purposes.

Rather forebodingly though, the *Nippon Times* suggested that "just as the development of aviation has claimed many victims it is quite likely that the development of atomic energy for peacetime purposes will do likewise from time to time."[36] The *Yomiuri Shimbun*-sponsored exhibition *Dare ni mo wakaru genshiryoku ten* (*Atomic Energy for Everyone Exhibition*) which was held later that month did not dwell on the dangers of atoms for peace but it was surprisingly up-front in terms of showing what the "bad" atom was capable of doing. Admission was free and with government support from the Ministry of Education, Science and Culture, Ministry of Health and Welfare, and the Ministry of Agriculture and Forestry. The venue was the seventh floor of the up-market Isetan Department Store in Shinjuku, Tokyo. Held from 12 to 22 August, just after the ninth anniversaries of the atomic bombs dropped on Hiroshima (6 August) and Nagasaki (9 August), the exhibition attracted large crowds. Coming less than six months after the Lucky Dragon Incident of 1 March, it was a timely exhibition which served to bring knowledge of atomic energy to the everyday person.[37]

A newspaper advertisement on the opening day of the exhibition depicted a mushroom cloud in the top left corner and a quick outline of major selling points of the exhibition. First up was the easy atomic physics classroom which included an *Alice in Atom-Land* diorama which allowed children to understand difficult theories "in a glance."[38] In the lower right corner of the advertisement, visitors were reminded that the Isetan was where "*okaimono wa suzushii*" ("Where Shopping is Cool and Refreshing"). The exhibition also boasted atomic bomb-related materials from Hiroshima and Nagasaki that were being exhibited in Tokyo for the first time.[39] Prior to the Lucky Dragon Incident that year, there had not been a widespread sense of what had happened in Hiroshima and Nagasaki. Now, there was a sense of triple victimhood.

Visitors were greeted by the bust of St Francis Xavier who worked in Japan as a Jesuit missionary from 1549 to 1551. That was all that remained of the stone statue that had stood at Urakami Cathedral in Nagasaki which was at Ground Zero, the epicentre of the bomb which exploded 500 metres above. Now in the exhibition, the statue's head was placed on a tall, cylindrical plinth at the entrance. Projecting from the plinth were four spokes which supported a Saturn ring-like band of signage bearing the word "Atomic" as if Xavier was a latter-day saint conveying his blessings to the modern world.[40] The head certainly added a humanistic element to the exhibition although its religious significance might not have been apparent to all. From Nagasaki also came a clock that stopped at 11:02 a.m. when the atomic bomb was dropped there, and a rosary chain belonging to a Nagasaki Christian. The melted tiles and remains of clothing from Hiroshima conveyed the awesome power of the bomb and its human cost.[41] Large photographs showing the terrible injuries that victims sustained made real their suffering.

As if to underline the continuing sacrifices that the Japanese were making, visitors were able to see for a short time the still slightly radioactive ship's wheel of the tuna fishing boat the Lucky Dragon No. 5,[42] items from the ship Shunkotsumaru that sailed around Bikini Atoll to survey the radioactivity of the sea water,[43] and the white ashes of nuclear fallout from Bikini Atoll that fell on the Lucky Dragon, as well as gloves and clothes of the crew. Technical equipment on display included radioactive contamination measurement instruments and a Wilson cloud chamber used to detect ionizing radiation.[44] Visitors could view a model of an atomic reactor, or crowd around a panorama showing what would happen if a hydrogen bomb fell on Shinjuku where the exhibition was taking place.[45] All this made for an atmosphere of considerable frission, but as the *Nippon Times* put it, it didn't stop jittery Tokyoites from jam-packing the exhibition.[46]

An *Alice in Atom-Land* diorama and Atomic Physics Classroom both sought to convey new knowledge about atomic energy to housewives and school children. The emphasis was on how atomic energy could impact on everyday life. There was obviously an understanding of the pivotal role of women in shaping attitudes to nuclear power (especially in light of the Lucky Dragon Incident earlier that year and nationwide panic regarding radioactive food). The department store was a congenial setting and a savvy way of targeting an audience that might not otherwise be attracted to what might appear to be a very technically oriented exhibition.[47]

Walt Disney's animated film *Alice in Wonderland* (1951) had been released in Japan in 1953,[48] the year before the exhibition, so the *Alice in Atom-Land* diorama drew on the popularity of the film and resonated with the characters that young visitors encountered on the screen. The diorama also seems to have been inspired by an illustrated children's book of the same name which was written by Noriyuki Sakikawa and Taiji Satō and published in 1954 by Chikuma Shobō. *Alice in Atom-Land*, the book, sought to introduce the world of science to primary school children. As per Lewis Carroll's original story *Alice's Adventures in Wonderland* (1865), Alice grows very small, but in Sakikawa and Satō's version, she travels through the molecular world, visits the museum of atom-land, and encounters particles smaller than an atom. The end of the story sees her searching for uranium.[49] In this way, Alice (and the young readers of the book) gained a positive view of the atomic world that adults could only dream about. Transferred to a diorama, *Alice in Atom-Land* was a useful way of balancing the frightening images of destruction elsewhere in the exhibition with an optimistic view of the future.[50] Alice's adventures mirrored the journey which Japan itself was undertaking in the atomic age. In the coming year, the Geological Survey Institute of the Ministry of International Trade and Industry was about to conduct basic surveys of what it considered to be the most promising uranium deposits.[51]

A newspaper advertisement also urged adult readers of the *Yomiuri Shimbun* to visit the exhibition. The Japanese people had experienced the atomic bombs dropped on Hiroshima and Nagasaki and also had been, recently, victims of the hydrogen bomb test at Bikini Atoll. They, of all people, should be interested in atomic energy. The narrative being implied was that the Japanese were entitled to exploit the peaceful and positive uses of atomic energy. The message was reinforced in the exhibition through easy-to-understand explanations, visual materials, models, actual materials and experiments.[52] Visitors would leave the exhibition having learnt what the atomic bomb and hydrogen bomb were; some sense of the history of the development of atomic physics leading up to the hydrogen bomb; and viewed documentary photographs of hydrogen bomb tests. Through such exhibitions, visitors would gain an awareness of the power of atomic weapons; learnt the basics of atomic physics; and come to an understanding of radioactivity, its effects on human beings and fish, and what was meant by radioactive rain. They would also have viewed, most likely for the first time, artefacts from Hiroshima and Nagasaki testifying to the power of the bomb. But on a more positive note, they would have seen how peaceful uses of atomic energy in agriculture, engineering, medicine and culture could benefit humankind. Visitors would have glimpsed a future of atomic powered trains, ships, airplanes and atomic power plants.[53]

How did visitors respond to the exhibition? Oketani Shigeo, an associate professor in engineering at the Tokyo Institute of Technology, reported his impressions which were published in the *Yomiuri Shimbun* on 14 August 1954, the day before the anniversary of Emperor Hirohito's radio broadcast to the Japanese people announcing that the government had been instructed to accept the terms of the Potsdam Declaration. Still just two years after the Allied Occupation of Japan had ended when the standard of living for many was not high, Oketani confessed to never having entered a department store. Entering on a hot summer's day, he was surprised at the strong air-conditioning in the store. He went up to the seventh floor and was surprised again to encounter a large number of people lining up to view the photographs and models on display. He didn't think so many would throng to the exhibition on the first day.[54]

The exhibition began with explanations of the atomic bomb and photographs and artefacts relating to the destruction at Hiroshima and Nagasaki. It reminded visitors of how terrible the bomb was and how a future war would lead to mass destruction. At the same time, the exhibition showed how atomic energy could be a powerful energy source. He thought that it was the first time that a model of an atomic reactor had been displayed in Japan. Although only small, it nevertheless accurately showed the workings of a reactor. He also noted the items from the Lucky Dragon and Shunkotsumaru, as well as other equipment which students from Tokyo University of Education were enthusiastically operating. Oketani had worried that the contents of the exhibition would be too difficult for members of the public to understand but he was pleased with the exhibition. He suggested that educational pamphlets would help, and that not only teachers and students but also everyone should visit the exhibition.[55]

In the final few days of the exhibition (20–22 August), there was a public program of guest lectures and film screenings in the Seventh Floor Hall at Isetan. On all three days, visitors could view a slide show of the effects of the atomic bomb on Hiroshima and Nagasaki; view films on atomic energy and its peaceful uses, courtesy of the American Cultural Centre; as well as view the film *Shi no hai* (*Ashes of Death*) courtesy of Puremia-Nikkatsu Films. On 21 August, there were two lectures: Dr Miyake Yasuo (Meteorological Research Institute) spoke on radioactive rain and Professor Kimura Kenjirō (University of Tokyo) delivered a lecture entitled "Uranium Resources." The following day, Professor Taketani Mituo (Rikkyō University) spoke on "The Power and Management of Atomic Energy." After the lectures, the speakers took questions from the public.[56]

Shōriki was pleased at the success of the exhibition and by late January 1955, arrangements were underway for a more fully fledged Atoms for Peace campaign. The initial plan was for Shōriki to use personnel and media already at his disposal and to rework CIA materials that the origins of the material were not evident and that they would suit the prevalent attitudes of the Japanese public and "general Japanese psychology."[57] What's more, emphasis would be placed on Russian refusal to participate in an all-out Atoms for Peace plan, assuming that this would appeal to the US government. It is fascinating to note that Shōriki was of the opinion that USIS sponsored Atoms for Peace programs were not likely to be readily accepted by the Japanese public. Requests were made for non-governmental and non-USIS material to be made available. This could include items of a commercial nature and include photographs, historical material, technical displays, working models of atomic power plants, educational films, slides, drawings and posters.[58] In this way, Shōriki was sensitive to anti-US feeling in Japan and trod carefully to ensure that there was not the perception that the Japanese were being fed American propaganda. We see, thus, the formation of what would constitute the contours of a Japanese nuclear culture that was shaped by sensitivity to the Japanese political landscape and its recent nuclear past.

On 27 January 1955, Sidney R. Yates (Democrat, Illinois) introduced a bill in the US House of Representatives to construct an atomic power reactor in Hiroshima. He did so, he said, because "I believe it is necessary that we place primary emphasis upon our intention that the atom shall be used in the cause of peace rather than for the destruction which war brings."[59] Yates confessed to the Speaker that he was not the originator of the proposal and that credit be given to AEC Commissioner Thomas E. Murray. The cost of the reactor would depend on the type and the power-potential. However, if a duplicate of the Duquesne reactor being built in Shippingport, Pennsylvania, was constructed that produced upwards of 60,000 kilowatts, it would cost approximately US$45 million with a possibility of reduced cost in future. Yates referred to existence of people like the Nobel Prize-winning physicist Yukawa Hideki as evidence that Japan had the human resources and technological capability to construct and maintain such a project. Yukawa was thus emblematic of Japan's scientific prowess, and it is no surprise that Shōriki would go to much effort to ensure that Yukawa was associated with the Atoms for Peace enterprise in Japan, most notably through membership of the Japan Atomic Energy Commission that would be established the year after in 1956.

As to the possibility of nuclear proliferation through the production of fissionable materials, Yates reassured the house that one reactor would certainly not shift the balance of power, and that "far from creating a military risk, this project would contribute to the military strength of Japan, and hence of all of free Asia."[60] Yates hoped that the reactor would be named the McMahon reactor after the late Senator Brien McMahon (Democrat, Connecticut) who spoke to the Senate on 2 February 1950, about his vision of atomic energy helping to feed the world:

> Perhaps through atomic power for industry and agriculture we can transform the deserts of Africa, Asia, and the Americas into blooming crop-producing acres, and the arid hills of the world into gardens. It is almost impossible to overestimate what all-out concentration upon atomic energy for peace might accomplish in terms of remaking and improving the physical environment of mankind.[61]

Members of the US Atomic Energy Commission were conscious of the danger that such images of blooming deserts might paint overly rosy pictures of the possibilities for atomic energy but Yates joined with Commissioner Murray in urging that an atomic power plant be built in Hiroshima as part of Eisenhower's Atoms for Peace plan.[62] Meanwhile, in the first half of 1955, other players became involved such as Nelson Rockefeller, Special Assistant to President Eisenhower. Rockefeller developed two proposals on international peaceful atomic development. The first involved a program of gifts of research reactors and the second proposed the transmission of classified information on power reactors to certain foreign countries. The Secretary of State, John Foster Dulles, noted in a letter to Rockefeller on 2 May 1955 that the first proposal might interfere with negotiations underway for the sale of research reactors to foreign countries. The second seemed to be along the lines that the National Security Council had already adopted as policy.[63]

The Geneva Atomic Fair

In July 1955, Lewis Strauss hosted the first International Conference on the Peaceful Uses of Atomic Energy, held under the auspices of the United Nations. Visitors were excited by the swimming pool reactor in the American exhibit but the conference also enabled visitors to compare their accomplishments with British, Soviet and French achievements in the development of nuclear power. The Soviets displayed a model of their 5000 kilowatt reactor that was already producing electricity, and the British showed a model of the 100,000 kilowatt Calder Hall reactor which was under construction by the government. It would boast almost twice the capacity of the American Shippingport reactor which would not produce electricity until late 1957. So while visitors left impressed by the American exhibit, it was clear that the British and the Soviets led the USA in civilian nuclear technology.[64]

Moorhead Patterson attended the conference and negotiated bilateral agreements. His position differed from Rockefeller's in that he favoured at least partial payment of the cost of the research reactors. President Eisenhower had recently offered to pay half. So increasingly, the agreements were more akin to commercial transactions rather than the result of a government-to-government relationship involving handouts. Indeed, the conference was likened to an atomic trade fair with British representatives seen as salesmen offering atomic power reactors like a product, with a promise of delivery in eighteen months. Even Sir John Cockcroft who headed the British delegation thought that such claims were fanciful but many could not help to be excited by the perception that an atomic-led revolution was occurring before their eyes.[65] The only problem was that as the nuclear historians Richard G. Hewlett and Jack M. Holl put it, "nuclear power was not yet ready for the marketplace."[66]

The optimistic image of nuclear power that was promoted at Geneva was carefully stage-managed. For example, the AEC excluded a presentation by the Indiana University geneticist H.J. Muller as his paper included references to the genetic effects of the atomic bombs dropped on Hiroshima and Nagasaki which were deemed inappropriate for a conference on the peaceful uses of atomic energy.[67]

Many of the US exhibits shown by companies at Geneva were displayed from 27 to 29 September 1955, at what was dubbed the first Atomic Trade Fair in the USA. It was held in conjunction with the third annual meeting of the US Atomic Industrial Forum at the Park-Sheraton Hotel, Washington, DC. Some seventy companies exhibited products ranging from complete reactors and instrumentation panels through to shielding materials and radioactive isotopes. Included among the foreign exhibits was a model of the British Calder Hall reactor power plant presented by the UK Atomic Energy Authority. The fair was designed to bring buyers and sellers together and ultimately Japan would show an interest in the Calder Hall reactor.[68]

Atoms for Peace Exhibition

In 1955, the recently remarried Frances Baker, now known as Frances Blakemore, supervised the design and construction of an *Atoms for Peace* exhibition in her capacity as chief exhibits officer for the USIS, based at the US Embassy in Tokyo. The exhibition was jointly sponsored by the US government and Shōriki's newspaper, *Yomiuri Shimbun*, which promoted it as providing the public with an opportunity to see "atoms for peace" with their very own eyes.[69] A photograph shows Frances Blakemore with the US Ambassador John M. Allison, inspecting Blakemore's preliminary design.[70] On 1 November 1955, the exhibition opened in Tokyo's Hibiya Park. *The New York Times* saw it as a test of Japan's readiness to accept "the atom as a friend" rather than as a destructive force. Scheduled to run for six weeks, the exhibition which was organised by the *Yomiuri Shimbun* and the USIS then moved to the major cities of Nagoya, Kyoto, Osaka, Hiroshima, Fukuoka, Sapporo and Sendai.

Mitsui Art Film Productions took footage of the exhibition's opening ceremony and produced an educational documentary for the USIS entitled *Power for Peace* which was released in 1956. The seventeen-minute colour film provides a valuable historical record of the exhibition, reflecting the hopes of the organizers and the messages that they sought to convey to the Japanese public regarding civilian atomic energy. It begins by showing a mechanical arm apparatus writing the film's title in English in the East Asian manner with brush and ink while another arm holds down the paper. These so-called magic hands were a popular attraction at the exhibition. They symbolized the coming together of East Asian culture and Western technology. The "magic hands" could be used by an operator to safely handle radioactive materials from a distance behind a transparent shield without direct contact. A closed-circuit television could be used to by an operator to view the operation so that even direct sight was not required.[71]

In the first scene, we see the word "ATOM" spelt out in large letters on a wall in front of which large numbers of Japanese are shown queueing to enter the exhibition. US Ambassador to Japan, John M. Allison, is shown conveying a message from President Eisenhower that the exhibit was a symbol of their "determination that the great power of the atom shall be dedicated to the arts of peace."[72] Shōriki, soon-to-be Chairman of Japan's Atomic Energy Commission, called the exhibit a curtain-raiser to the atomic age. Chief Cabinet Secretary Nemoto Ryutarō read a message from Prime Minister Hatoyama Ichirō praising atomic science and welcoming the exhibition as an educational initiative.[73]

The display was seen as being comprehensive in that it showed every type of reactor that had been devised up until that time. Exhibits included a model of a medical reactor being built at the University of California, two swimming pool reactors, and a miniature model of a futuristic, atomic powered train that no doubt drew parallels with the model train that the American Commodore Matthew C. Perry brought to Japan in 1854 when the nation was force to open its ports to the West.[74]

The exhibition was divided into fourteen parts.[75] The first profiled ten scientists including Japan's Yukawa Hideki, as pioneers of atomic energy. The documentary of the exhibition explained how Yukawa had "provided new facts about forces in the atom's nucleus with his meson prediction theory."[76] In this way, Japan gained some sense of ownership of its development. After dwelling on this history, visitors then learnt some basic facts regarding atomic energy and viewed a film on the peaceful uses of atomic energy. They then inspected a model of a reactor and viewed a display of educational wall panels. There was very much a sense of progression with subsequent sections devoted to a model of a Graphite Moderated Natural

Uranium Reactor from Oak Ridge National Laboratory in the USA and explanations of a nuclear chain reaction. The ORNL Graphite Reactor was even in the USA, seen as a useful way to train personnel in understanding how to operate a reactor, the impact of radiation and radiation hazards.[77]

Visitors were able to learn of the importance of radioisotopes in industry, medicine and agriculture, and view a model laboratory. They learnt of how radiation could assist in the preservation of food products, and how there were already efforts being made to promote knowledge of atomic energy in education and research. Looking to the future, visitors could view a diorama showing atomic-powered trains and an atomic power plant. Visitors then entered a room devoted to a display on uranium mining which would supply the fuel for the plant. Those who want to learn more could then spend time browsing through publications before exiting the exhibition. It was a carefully thought-out exhibition which, in contrast to *Atomic Energy for Everyone* that had been held only a few months earlier at Isetan, unabashedly showed the atom in a highly positive light.

The exhibition explained how researchers were hoping to discover reactor designs and techniques best suited to the production of electricity. Through a combination of animation, artist-drawn images, photographs, clever lighting effects and models, the exhibition helped make "nuclear power" visible to the public by explaining the process of uranium fission, how it was discovered, how it could be safely controlled, and how nuclear fission could be harnessed to produce electricity. As the documentary of the exhibition assured viewers, this would provide "cities and towns with light and heat at a fraction of the consumption of conventional fuels."[78]

Just two weeks after the opening of the exhibition, the Japanese Ambassador to Washington, DC, Iguchi Sadao, signed the Japan-US Atomic Energy Cooperation Agreement, with AEC Chairman Lewis L. Strauss and William Sebald, Deputy Assistant Secretary of State for the Far East signing for the USA. The five-year agreement included the lease of enriched uranium as fuel for experimental reactors, and assistance with the construction and operation of the reactor. Japan planned to purchase a boiling water reactor (developed by Idaho National Laboratory and General Electric), to be followed by a CP-5 type reactor (which used enriched uranium as fuel) and a swimming pool reactor.[79] The *Nippon Times* viewed the signing of the agreement as perhaps the most important, single event in Japan's post-war history, the first step into the Atomic Age. The semi-governmental Japan Atomic Energy Research Institute was scheduled to start operating by the end of the month and the newspaper noted that "the nation is finally recovering from 'nucleophobia,' which has been largely responsible for Japan's backwardness in peaceful atomic energy development." The newspaper diagnosed this as a "nervousness over atomic energy" which was deemed "natural for a people who were the first to be exposed to the atomic bomb." However, making clear where its sympathies lay, it suggested that "it ranks with primitive superstition to allow this to stand in the way of the benefits we may reap from the atom." In this way, the newspaper and other commentators medicalized nuclear fear to suggest that why it was understandable: such fears were folly.[80] The editorial also acknowledged that nuclear fear had not disappeared overnight. Support for increasing investment in nuclear power development would only grow, it noted, if people were able to link it to an improved standard of living. In this respect, the *Nippon Times* saw the US *Atoms for Peace* exhibition as being invaluable, helping to allay fears of nuclear power and promoting the idea that it was safe and the key to Japan's future.[81]

The exhibition closed on 12 December 1955 and it was estimated that 367,669 people had seen it, making for a daily average of 8754, a world record attendance for a USIS exhibit of this nature. The positive response to the exhibition was seen as a barometer of public understanding of the peaceful uses of atomic energy.[82] Approximately 60 per cent of visitors had purchased a USIS-produced information pamphlet and many had read articles, especially in the *Yomiuri Shimbun* on the exhibition.[83] The only place that came close to attracting the response seen in Tokyo was that in Berlin where over six weeks some 220,000 visitors attended a comparable exhibition.[84] The exhibition toured within Japan for two years, visiting seven other Japanese cities including Hiroshima.

Shōriki also personally benefited by being associated with the exhibition. He had now become Minister of State and Head of the Hokkaido Development Agency, as well as Deputy Chair of the Science and Technology Administration Council (STAC)[85] which was headed by Prime Minister Hatoyama Ichirō. Atomic energy had fuelled his rise to power and he entertained hopes of one day leading the nation himself. He shared his thoughts with the progressive *Asahi Shimbun*, a rival to his own more conservative newspaper, some five weeks after the exhibition had closed.[86] He suggested that Japan had lost the war due to its reluctance to introduce foreign things into the country and that it still was clueless in that regard. It was no good shutting itself away in a small box. He thought that it was important to send people overseas to learn. He himself had only learnt about what atomic energy had to offer in 1954.

One milestone of the past year was the *Yomiuri Shimbun*-sponsored visit to Japan in May 1955 by the General Dynamics President John Jay Hopkins who was accompanied by Dr Lawrence R. Hafstad (atomic research director, Chase Manhattan Bank, formerly director of atomic reactor research, US AEC) and Nobel Prize-winning physicist Professor Ernest O. Lawrence (University of California, Berkeley). During their visit, they were reported as having made three key points. Firstly, they claimed that there was no lingering radiation after an atomic bomb explosion, only during the explosion itself. Secondly, despite fears of strings being attached to any US support, the only condition on an offer of uranium to Japan was that the fissionable material not be used for military purposes. And thirdly, the USA was seeking to plant the seed of atomic energy in countries that needed it the most. Hopkins spoke of how atomic power could propel boats, produce radioisotopes, help preserve food and assist in agricultural production.[87] On 6 June, shortly after his return to the USA, Hopkins reported that a bilateral agreement was being negotiated in Washington, DC, through which an experimental research reactor program could be commenced in Japan within six months. If all went well, research reactors could be exported to Japan as part of a twenty-five-year program for the development of peaceful atomic energy in Asia. Hopkins foresaw that Japan would have some fifteen experimental reactors within the next five years. Japan would initially have to outlay some US$1,000,000 and spend approximately $25,000,000 over the next five or six years.[88] Thus, while Japan would have to bear much of the cost itself, the USA would share know-how and materials.[89]

The net result of all that happened in the year was that whereas at the beginning of 1955, there were many Japanese who were opposed to the development of atomic energy in Japan, opposition had seemed to have largely fallen away. Newspapers and visits by people such as Hopkins had helped to educate the general public about peaceful uses of atomic energy. This, he felt was very important and went to the very core of the mission of newspapers in society.[90]

As for who would fund the development of atomic energy in Japan, Shōriki felt that the private sector could not do it alone. The state had to fund and manage things and they would also look to the USA for funding. The journalist was concerned that such funding would come with strings attached, but Shōriki was adamant that if the USA placed difficult conditions, Japan could simply say no and turn to Japanese-made reactors instead. He told the *Asahi Shimbun* reporter that it was planned to purchase an experimental boiling water reactor (BWR) from the USA in 1956 and a CP-5 (Chicago Pile-5) research reactor from the USA the year after. It was thus crucial to have the technical staff and infrastructure in place as soon as possible in Japan in order to introduce such technology. As for who would be members of the Japan Atomic Energy Commission that he was set to chair from January 1956, he very much hoped that Nobel laureate Yukawa Hideki would agree to join. Yukawa would give the Commission credibility in the eyes of the world.[91]

Conclusion

We have seen in this chapter how Americans, even before Eisenhower's speech in 1953, were contemplating the symbolic value of harnessing atomic energy for peaceful purposes and how Japan, not least Hiroshima, would be an appropriate location for an American-sponsored power reactor. Shōriki was alert to these proposals and saw atomic energy as an opportunity to promote his political career and help his country at the same time. The speeches and rhetoric in America directly informed the vision of atoms for peace in Japan and nicely correlated with sentiment by Taketani and others that Japan were entitled to or were owed this. Photographs and film were crucial in making these hopes as well as concerns about atomic energy visible. Chapter 5 focuses on the impact of the 1954 Lucky Dragon Incident and the story of the controversial *Family of Man* photographic exhibition at Tokyo's Takashimaya Department Store in 1956. Chapter 6 examines several films that explore how moving images were used to allay and express nuclear fears especially in the wake of the Lucky Dragon Incident.

Notes

1. *US Congressional Record*, Senate, 2 Feb. 1950, 1340.
2. Taketani, Mituo, "Nihon no genshiryoku kenkyū no hōkō" ("The Direction of Atomic Energy Research"), *Kaizō* (*Reconstruction*) 33, no. 17 (Nov. 1952): 70–72. Partially reprinted in Taketani, Mituo, *Genshiryoku to kagakusha* (*Atomic Energy and Scientists*), Taketani Mituo chosakushū 2 (Collected Works of Taketani Mituo), Vol. 2 (Tokyo: Keisō Shobō, 1968), 471. Translation based on Hitoshi Yoshioka, "Nuclear Power Research and the Scientists' Role" in *A Social History of Science and Technology in Contemporary Japan, Vol. 2: Road to Self-Reliance, 1952–1959*, eds.

Shigeru Nakayama with Kunio Gotō and Hitoshi Yoshioka (Melbourne: Trans Pacific Press, 2005), 104–24, esp. 112. Some corrections to the translation have been made.

3. Edward F. Ryan, "U.S. to Build A-Plant for Industry," *Washington Post*, 23 Oct. 1953, 1–2, esp. 1; Richard G. Hewlett and Jack M. Holl, *Atoms for Peace and War, 1953–1961: Eisenhower and the Atomic Energy Commission* (Berkeley: University of California Press, 1989), 194.
4. Ryan, "U.S. to Build A-Plant for Industry," esp. 2.
5. Ryan, "U.S. to Build A-Plant for Industry," 2.
6. President Dwight D. Eisenhower, "Atoms for Peace" Address Before the General Assembly of the United Nations on Peaceful Uses of Atomic Energy, New York City, 8 Dec. 1953, Eisenhower Presidential Library, accessed 2 Feb., 2017, https://www.eisenhower.archives.gov/all_about_ike/speeches/atoms_for_peace.pdf, 5.
7. Eisenhower, "Atoms for Peace" Address, 7.
8. US Delegation to the Eighth General Assembly, "Summary of Reactions to President Eisenhower's Speech Before the General Assembly, December 8, 1953," Confidential Report, 9 Dec. 1953, Digital National Security Archive collection: Nuclear Non-proliferation.
9. US Delegation to the Eighth General Assembly, "Summary of Reactions to President Eisenhower's Speech Before the General Assembly, December 8, 1953," 8.
10. Shane J. Maddock, *Nuclear Apartheid: The Quest for American Atomic Supremacy from World War II to the Present* (Chapel Hill: University of North Carolina Press, 2010), 79.
11. Gerard C. Smith, *Disarming Diplomat: The Memoirs of Gerard C. Smith, Arms Control Negotiator*, foreword by Henry Owen, introduction by Kenneth W. Thompson (Lanham, Maryland: Madison Books, 1996), 29.
12. US Atomic Energy Commission General Advisory Committee 1954, *Thirty-Eighth Meeting of the General Advisory Committee to the U.S. Atomic Energy Commission, January 6, 7, and 8, 1954, Washington, D.C.*, 13. Digital National Security Archive: Nuclear Non-proliferation.
13. US Atomic Energy Commission General Advisory Committee 1954, 27. Digital National Security Archive: Nuclear Non-proliferation.
14. US CIA, Office of National Estimates, "Political and Psychological Effects of a US Program for Cooperation with Other Nations in the Peaceful Uses of Atomic Energy," Draft Memorandum for the Director of Central Intelligence, 9 Aug. 1954, accessed 2 Feb. 2017, https://www.cia.gov/library/readingroom/document/cia-rdp79r00904a000200010018-9
15. US Central Intelligence Agency, Office of National Estimates, "Political and Psychological Effects," 3.
16. US CIA, "Political and Psychological Effects," 4.

17. US CIA, "Political and Psychological Effects," 11.
18. Thomas E. Murray, "Atomic Energy for Peace," *US Congressional Record*, House of Representatives, 27 Jan. 1955, Appendix, 875–877, esp. 876.
19. Murray, "Atomic Energy for Peace," 877.
20. Murray, "Atomic Energy for Peace;" Stanley Levey, "Nuclear Reactor Urged for Japan," *New York Times*, 22 Sept. 1954, 14.
21. Murray, "Atomic Energy for Peace," 876.
22. Murray, "Atomic Energy for Peace," 877.
23. Smith, *Disarming Diplomat*, 2, 25–27.
24. "A Reactor for Japan," *Washington Post*, 23 Sept. 1954, 18.
25. "Memorandum of Conversation, by the Ambassador to Japan (Allison)," 9 Nov. 1954, *Foreign Relations of the United States, 1952–1954, China and* Japan, Vol. XIV, part 2, document no. 825, accessed 6 Feb. 2017, https://history.state.gov/historicaldocuments/frus1952-54v14p2/d825
26. Roland Sawyer, "Mr Murray's Proposal," *Christian Science Monitor*, 24 Sept. 1954, *US Congressional Record*, House of Representatives, 27 Jan. 1955, Appendix, 877.
27. John Jay Hopkins, "An 'Atomic Marshall Plan': The Development of International Atomic Energy under Leadership of American Industry," *Vital Speeches of the Day* 21, no. 8 (1 Feb. 1955), 1007–1012; see also John Jay Hopkins, "The Development of International Atomic Energy under the Leadership of American Industry," *US Congressional Record*, House of Representatives, 27 Jan. 1955, Appendix, 878–881.
28. Hopkins, "The Development of International Atomic Energy," 880.
29. Hopkins, "The Development of International Atomic Energy," 880.
30. "On Spreading the Atom Around," *Los Angeles Times*, 19 Jan. 1955, A4.
31. US CIA, Memorandum for Chief, Information Coordination Division, "Exploitation of Atoms for Peace Program in Japan," 17 Jan. 1955, Special Collection, Nazi War Crimes Disclosure Act, Shoriki Matsutaro Vol. 1, no. 32, document no. 519cd821993294098d516e57. See also: US CIA, Memoranda for Chief, Security Office, Special Security Division, "Proposed Trip to Japan by Mr. John Jay Hopkins and Mr. Will S. Halstead," 24 Jan. 1955, Special Collection, Nazi War Crimes Disclosure Act, Shoriki Matsutaro Vol. 1, no. 33, document no. 519cd821993294098d516e70.
32. Day Inoshita, "A-Bomb City Prays Peace," *Nippon Times*, 7 Aug. 1954, 3.
33. "A-Bomb City Slates Rites," *Nippon Times*, 6 Aug. 1954, 3.
34. "First Atomic Bomb," *Nippon Times,* 6 Aug. 1954, 8.
35. Day Inoshita, "A-Bomb City Prays Peace," *Nippon Times*, 7 Aug. 1954, 3.
36. "First Atomic Bomb."
37. "*Dare ni mo wakaru genshiryoku ten*" ("*Atomic Energy for Everyone Exhibition*"), *Yomiuri Shimbun*, 9 Aug. 1954, 7.

38. "*Dare ni mo wakaru genshiryoku ten*," advertisement, *Yomiuri Shimbun*, 12 Aug. 1954, 5.
39. "Tōkyō ni saigen suru: kyūnen mae no Hiroshima Nagasaki" ("Hiroshima and Nagasaki of Nine Years Ago, Reappears in Tokyo"), *Yomiuri Shimbun*, 8 Aug. 1954, evening edition, 4.
40. Illustrated in "'Genshiryokuten' hiraku" ("'The Atomic Energy Exhibiton' Opens"), *Yomiuri Shimbun*, 12 Aug. 1954, evening edition, 3.
41. "Bikini no hai mo shokōkai" ("First Public Display of Nuclear Fallout from Bikini Atoll"), *Yomiuri Shimbun*, 12 Aug. 1954, 7.
42. Illustrated in "Fukuryūmaru no kaji" ("The Lucky Dragon Ship's Wheel"), *Yomiuri Shimbun*, 13 Aug. 1954, 6; see also "Kuzuoreta seizō no kubi" ("The Head of a Venerated Fallen Statue"), *Yomiuri Shimbun*, 14 Aug. 1954, 6.
43. H.K. Yoshihara, "The Dawn of Radiochemistry in Japan," *Radiochimica Acta* 100 (2012): 523–527, esp. 526.
44. "Dare ni mo wakaru genshiryoku ten," advertisement, *Yomiuri Shimbun*, 12 Aug. 1954, 5.
45. "Suibaku ga Shinjuku ni ochitara" ("If a Hydrogen Bomb was Dropped on Shinjuku"), *Yomiuri Shimbun*, 18 Aug. 1954, 6.
46. "Jittery Tokyoites Jampack Atomic Energy Exhibition," *Nippon Times*, 14 Aug. 1954, 3.
47. "Bikini no hai mo shokōkai" ("First Public Display of Nuclear Fallout from Bikini Atoll"), *Yomiuri Shimbun*, 12 Aug. 1954, 7.
48. Jonathan Clements, *Anime: A History* (London: Palgrave Macmillan on behalf of British Film Institute, 2013), 83.
49. Noriyuki Sakikawa and Taiji Satō, *Genshi no kuni no Arisu* (*Alice in Atom-Land*) (Tokyo: Chikuma Shobō, 1954).
50. The diorama may have also been inspired by the Exhibition of Science at the 1951 Festival of Britain. See Sophie Forgan, "Atoms in Wonderland," *History and Technology* 19, no. 3 (2003): 177–196.
51. "Local Uranium Mining to Move Into High Gear," *Nippon Times*, 4 Jan. 1956, 3.
52. "Dare ni mo wakaru genshiryoku ten," advertisement, *Yomiuri Shimbun*, 12 Aug. 1954, 5.
53. "*Dare ni mo wakaru genshiryoku ten*" ("*Atomic Energy for Everyone Exhibition*"), *Yomiuri Shimbun*, 9 Aug. 1954, 7.
54. Shigeo Oketani, "'Genshiryokuten' o mite" ("Seeing the 'Atomic Energy Exhibition'"), *Yomiuri Shimbun*, 14 Aug. 1954, 8.
55. Shigeo Oketani, "'Genshiryokuten' o mite" ("Seeing the 'Atomic Energy Exhibition'"), *Yomiuri Shimbun*, 14 Aug. 1954, 8.

56. "*Dare ni mo wakaru genshiryoku ten*: Tokubetsu kōen to eiga no kai" ("*Atomic Energy for Everyone*: Special Lectures and Films"), *Yomiuri Shimbun*, 20 Aug. 1954, 7.
57. US CIA, Memoranda for Chief, Information Coordination Division, "Exploitation of Atoms for Peace Program in Japan," 28 Jan. 1955, Special Collection, Nazi War Crimes Disclosure Act, Shoriki Matsutaro Vol. 1, no. 35, document no. 519cd821993294098d516e5a.
58. US CIA, "Exploitation of Atoms for Peace Program in Japan," 28 Jan. 1955.
59. *US Congressional Record*, House of Representatives, 27 Jan. 1955, 873.
60. *US Congressional Record*, House of Representatives, 27 Jan. 1955, 875.
61. *US Congressional Record*, House of Representatives, 27 Jan. 1955, cited on 875.
62. International News Service, "Belgium and Japan Seek 1st 'A-for-Peace' Power," *Washington Post*, 15 Feb. 1955, 5.
63. US Department of State, *Foreign Relations of the United States, 1955–1957, Regulation of Armaments; Atomic Energy*, Vol. XX, accessed 10 Feb. 2017, https://history.state.gov/historicaldocuments/frus1955-57v20/d21; see also US CIA, General CIA Records, "Letter to the Honorable Nelson Rockefeller from Richard M. Bissell, Jr.," 19 April 1955, document no. CIA-RDP80B01676R004300080013–5, accessed 10 Feb. 2017, https://www.cia.gov/library/readingroom/document/cia-rdp80b01676r004300080013-5
64. Richard Pfau, *No Sacrifice Too Great: The Life of Lewis L.* Strauss (Charlottesville: University Press of Virginia, 1984), 192–94.
65. Marquis Childs, "Atomic Revolution is Theme at Geneva," *Washington Post*, 16 Aug. 1955, 18.
66. Hewlett and Holl, *Atoms for Peace and War, 1953–1961*, 208.
67. Eugene Rabinowitch, "Genetics in Geneva," *Bulletin of the Atomic Scientists* (Nov. 1955): 314–16, 343.
68. Our US Industrial Correspondent, "Atomic Trade Fair," *Financial Times*, 14 Sept., 1955, 2.
69. "Me de miru 'genshiryoku heiwa riyō'" ("Seeing 'Atoms for Peace with One's Own Eyes"), *Yomiuri Shimbun*, 8 May, 1955, 3.
70. Michiyo Morioka, *An American Artist in Tokyo: Frances Blakemore, 1906–1997* (Seattle: The Blakemore Foundation, 2007), 126–127.
71. US Government, USIS, *Power for Peace*, seventeen minute colour film (Tokyo: Mitsui Art Film Productions, 1956). A photograph of the "magic hands" can be seen in Ran Zwigenberg, *Hiroshima: The Origins of Global Memory Culture* (Cambridge: Cambridge University Press, 2014), 120.
72. US Government, USIS, *Power for Peace*, seventeen minute colour film (Tokyo: Mitsui Art Film Productions, 1956).

73. US Government, USIS, *Power for Peace*, seventeen minute colour film (Tokyo: Mitsui Art Film Productions, 1956).
74. Associated Press, "Atom-Peace Show Opens Here Today," *Nippon Times*, 1 Nov. 1955, 1.
75. See Tetsuo Arima, *Genpatsu, Shōriki, CIA* (*Atomic Power, Shōriki and the CIA*) (Tokyo: Shinchosha, 2008), 119.
76. US Government, USIS, *Power for Peace*, seventeen minute colour film (Tokyo: Mitsui Art Film Productions, 1956).
77. C.D. Cagle, *The Oak Ridge National Laboratory Graphite Reactor*, ORNL Central Files No. 53-12-126 (Oak Ridge, Tennessee: ORNL, c. 1957), http://web.ornl.gov/info/reports/1957/3445605702068.pdf
78. US Government, USIS, *Power for Peace*, seventeen minute colour film (Tokyo: Mitsui Art Film Productions, 1956).
79. "Japan-U.S. A-Pact to be Inked Today," *Nippon Times*, 15 Nov. 1955, 1.
80. "Japan Enters the Atomic Age," editorial, *Nippon Times*, 16 Nov. 1955, 8.
81. "Japan Enters the Atomic Age," editorial, *Nippon Times*, 16 Nov. 1955, 8.
82. "Daihankyō o yonda genshriyoku heiwa riyō haku" ("Atoms for Peace Exhibition Met with Big Response"), *Yomiuri Shimbun*, 13 Dec. 1955, 8.
83. See Tetsuo Arima, *Genpatsu, Shōriki, CIA* (*Atomic Power, Shōriki and the CIA*) (Tokyo: Shinchosha, 2008), 120–21.
84. "Daihankyō o yonda genshriyoku heiwa riyō haku" ("Atoms for Peace Exhibition Met with Big Response"), *Yomiuri Shimbun*, 13 Dec. 1955, 8.
85. "Genshiryoku ni zenryoku tsukusu" ("Making An All-Out Effort to Promote Atomic Energy"), *Yomiuri Shimbun*, 26 Nov. 1955, 1.
86. "Nani o kangae, nani o suru?" ("What are you thinking? What do you plan to do?), *Asahi Shimbun*, 21 Dec. 1955, 1.
87. Associated Press, "Atom Power Prospects Detailed for Japanese," *Christian Science Monitor*, 17 May, 1955, 6.
88. "U.S. to Aid Japan on Atomic Power," *New York Times*, 7 June 1955, 47, 49.
89. Robert C. Cowen, "Interview with John Jay Hopkins," *Christian Science Monitor*, 21 July, 1955, 9.
90. "Nani o kangae, nani o suru?" ("What are you thinking? What do you plan to do?), *Asahi Shimbun*, 21 Dec. 1955, 1.
91. "Nani o kangae, nani o suru?" ("What are you thinking? What do you plan to do?), *Asahi Shimbun*, 21 Dec. 1955, 1.

Bibliography

Arima, Tetsuo. *Genpatsu, Shōriki, CIA (Atomic Power, Shōriki and the CIA)*. Tokyo: Shinchosha, 2008.

Asahi Shimbun, 21 Dec. 1955.

Cagle, C.D. *The Oak Ridge National Laboratory Graphite Reactor*, ORNL Central Files No. 53-12-126. Oak Ridge, Tennessee: ORNL, c. 1957. http://web.ornl.gov/info/reports/1957/3445605702068.pdf.

Christian Science Monitor, 24 Sept. 1954, 17 May, 1955, 21 July, 1955.

Clements, Jonathan. *Anime: A History*. London: Palgrave Macmillan on behalf of British Film Institute, 2013.

Eisenhower, President Dwight D. "Atoms for Peace" Address Before the General Assembly of the United Nations on Peaceful Uses of Atomic Energy, New York City, 8 Dec. 1953, Eisenhower Presidential Library. Accessed 2 Feb., 2017. https://www.eisenhower.archives.gov/all_about_ike/speeches/atoms_for_peace.pdf.

Financial Times, 14 Sept., 1955.

Forgan, Sophie. "Atoms in Wonderland." *History and Technology* 19, no. 3 (2003): 177–196.

Hewlett, Richard G. and Jack M. Holl, *Atoms for Peace and War, 1953–1961: Eisenhower and the Atomic Energy Commission*. Berkeley: University of California Press, 1989.

Hopkins, John Jay. "An 'Atomic Marshall Plan': The Development of International Atomic Energy under Leadership of American Industry." *Vital Speeches of the Day* 21, no. 8 (1 Feb. 1955a), 1007–1012.

Hopkins, John Jay. "The Development of International Atomic Energy under the Leadership of American Industry." *US Congressional Record*, House of Representatives, 27 Jan. 1955b, Appendix, 878–881.

Los Angeles Times, 19 Jan. 1955.

Maddock, Shane J. *Nuclear Apartheid: The Quest for American Atomic Supremacy from World War II to the Present*. Chapel Hill: University of North Carolina Press, 2010.

"Memorandum of Conversation, by the Ambassador to Japan (Allison)", 9 Nov. 1954, *Foreign Relations of the United States, 1952–1954, China and* Japan, Vol. XIV, part 2, document no. 825. Accessed 6 Feb. 2017. https://history.state.gov/historicaldocuments/frus1952-54v14p2/d825.

Morioka, Michiyo. *An American Artist in Tokyo: Frances Blakemore, 1906–1997*. Seattle: The Blakemore Foundation, 2007.

Murray, Thomas E. "Atomic Energy for Peace," *US Congressional Record*, House of Representatives, 27 Jan. 1955, Appendix, 875–877.

New York Times, 22 Sept. 1954, 7 June 1955.

Nippon Times, 6–14 Aug. 1954, 1–16 Nov. 1955, 4 Jan. 1956.

Pfau, Richard. *No Sacrifice Too Great: The Life of Lewis L.* Strauss. Charlottesville: University Press of Virginia, 1984.

Rabinowitch, Eugene. "Genetics in Geneva." *Bulletin of the Atomic Scientists* (Nov. 1955): 314–16, 343.

Sakikawa, Noriyuki and Taiji Satō, *Genshi no kuni no Arisu* (*Alice in Atom-Land*). Tokyo: Chikuma Shobō, 1954.

Smith, Gerard C. *Disarming Diplomat: The Memoirs of Gerard C. Smith, Arms Control Negotiator*. Foreword by Henry Owen, introduction by Kenneth W. Thompson. Lanham, Maryland: Madison Books, 1996.

Taketani, Mituo, "Nihon no genshiryoku kenkyū no hōkō" ("The Direction of Atomic Energy Research"), *Kaizō* (*Reconstruction*) 33, no. 17 (Nov. 1952): 70–72.

Taketani, Mituo, *Genshiryoku to kagakusha (Atomic Energy and Scientists)*, Taketani Mituo chosakushū 2 (Collected Works of Taketani Mituo), Vol. 2. Tokyo: Keisō Shobō, 1968.

US Atomic Energy Commission General Advisory Committee 1954, *Thirty-Eighth Meeting of the General Advisory Committee to the U.S. Atomic Energy Commission, January 6, 7, and 8, 1954, Washington, D.C.*, 13. Digital National Security Archive: Nuclear Non-proliferation.

US CIA, General CIA Records.

US Congressional Record, Senate, 2 Feb. 1950, House of Representatives, 27 Jan. 1955.

US Delegation to the Eighth General Assembly, "Summary of Reactions to President Eisenhower's Speech Before the General Assembly, December 8, 1953," Confidential Report, 9 Dec. 1953, Digital National Security Archive collection: Nuclear Non-proliferation.

US Department of State, *Foreign Relations of the United States, 1955–1957, Regulation of Armaments; Atomic Energy*, Vol. XX. Accessed 10 Feb. 2017. https://history.state.gov/historicaldocuments/frus1955-57v20/d21.

US Government, USIS, *Power for Peace*, seventeen minute colour film. Tokyo: Mitsui Art Film Productions, 1956.

Washington Post, 23 Oct. 1953, 23 Sept. 1954, 15 Feb. 1955, 16 Aug. 1955.

Yomiuri Shimbun, 8–20 Aug. 1954, 8 May, 1955, 26 Nov. 1955, 13 Dec. 1955.

Yoshihara, H.K. "The Dawn of Radiochemistry in Japan." *Radiochimica Acta* 100 (2012): 523–527.

Yoshioka, Hitoshi. "Nuclear Power Research and the Scientists' Role" in *A Social History of Science and Technology in Contemporary Japan, Vol. 2: Road to Self-Reliance, 1952–1959*, edited by Shigeru Nakayama with Kunio Gotō and Hitoshi Yoshioka, 104–124. Melbourne: Trans Pacific Press, 2005.

Zwigenberg, Ran. *Hiroshima: The Origins of Global Memory Culture*. Cambridge: Cambridge University Press, 2014.

CHAPTER 5

Nuclear Testing in the Pacific: The Lucky Dragon Incident and the *Family of Man*

Introduction

We saw in Chap. 3 how the Hiroshima panels painted by Akamatsu Toshiko (also known as Maruki Toshi) and her husband Maruki Iri were displayed at exhibitions throughout Japan and later toured the world. Despite this, the exhibitions did not translate into a nationwide movement until March 1954 when a Japanese fishing boat and its crew were exposed to radioactive fallout from an American hydrogen bomb test at Bikini Atoll in the South Pacific. The first part of this chapter examines that incident and the perception that the Japanese had fallen victim a third time to an American-made nuclear weapon. We examine how articles in the illustrated magazine *Asahi Gurafu* (*Asahi Graph*) were important in keeping the Japanese public informed about developments in the controversy using photographs to give them a sense of how the threat posed by nuclear testing could potentially affect them all. The American-initiated *Family of Man* photographic exhibition that subsequently toured Japan will be discussed in the second part of this chapter. It can be viewed as an attempt to change the narrative and to remind the Japanese of the universality of life and suffering.

M. Low, *Visualizing Nuclear Power in Japan*, Palgrave Studies in the History of Science and Technology,
https://doi.org/10.1007/978-3-030-47198-9_5

Voyage of the Lucky Dragon

On 1 March 1954, the *Nippon Times* English language newspaper reported on the forthcoming atomic bomb and hydrogen bomb tests that were to be conducted at the atolls of Eniwetok and Bikini in the Marshall Islands.[1] While informed readers of the international community in Japan were thus aware of the nuclear tests, detailed information had not been conveyed to the twenty-three crewmen of the tuna trawler Daigo Fukuryū-maru (The Lucky Dragon, No. 5) who were fishing in the Pacific. In the Foreword to Ralph S. Lapp's account of the voyage, Pearl S. Buck likened the story of the humble fishermen to a Greek tragedy.[2]

The story begins with the Lucky Dragon leaving from the port of Yaizu in Shizuoka prefecture where it was based and first heading to Midway Island near Hawaii. It then ventured south to the Marshall Islands. The crew knew that the USA had conducted atomic bomb tests back in 1946 and weren't aware of new tests in 1954. The fishing boat was 70–100 miles away from Bikini Atoll when a hydrogen bomb exploded there on 1 March. One of the crewmen, Suzuki Shinzō, remarked that the sun seemed to rise in the West, and shortly after there was a massive blast. A soundwave enveloped the trawler and some two hours later, white radioactive ash rained down upon them. The crew speculated that it was a "*pika-don*" (atomic bomb explosion).[3] The boat returned to the port of Yaizu on the morning of 14 March and five members of the crew were hospitalized later that day. Four more entered hospitals a few days later.

The *Yomiuri Shimbun* broke the news of the Lucky Dragon Incident in a late morning edition of the newspaper on 16 March 1954.[4] As soon as the headlines hit the street, other newspapers also began providing daily reports on the controversy which blew up and led to the Japanese Ministry of Foreign Affairs contacting the US Embassy in Tokyo and the Japanese Embassy in Washington, DC. The front page of the *Nippon Times* on 17 March featured two photographs which summed up Japanese fears.[5] One photo showed Shimizu Kentarō of the University of Tokyo examining deckhand Masuda Sanjirō who was suffering from secondary radiation burns from direct exposure to the radioactive ash. A second photo showed three members of the Scientific Research Institute (formerly known as Riken) armed with Geiger counters, inspecting tuna from the Lucky Dragon at the Tokyo Central Fish Market. The caption read "Hot Fish," alluding to how the fish had been contaminated by radioactivity. The scientists warned of the danger of prolonged human exposure to the fish.

The fish was subsequently buried in a vacant lot along with other fish in storage. Unfortunately, much of the Lucky Dragon cargo had already been shipped to locations in Tokyo and Osaka before the fish could be inspected, but the article assured readers that the fish had been seized before they could go on sale to members of the public.[6]

Meanwhile, Yaizu-sourced fish were examined on 16 March at the Osaka fish market by the biophysicist Nishiwaki Yasushi from Osaka City Medical College (now known as the Osaka City University). His Geiger counter detected radiation and the instrument emitted a sound which made the gathered crowd think that the fish were crying. Some of the contaminated fish had already been sold and eaten by some hundred people. The ensuing fear and evening newspaper reports result in people not buying fish, a staple food in the Japanese diet. On that same day, Kimura Kenjirō, professor in the Department of Chemistry at the University of Tokyo, examined some of the pale radioactive ash from the Lucky Dragon and his laboratory conducted an analysis in the hope of helping ailing crew members.[7] While most of the attention was focused on the Lucky Dragon, it was also reported that eight other tuner trawlers that had been fishing in the South Pacific had shown slight signs of radioactivity that was not harmful. The government also sought to assure consumers that all fish would be inspected and encouraged housewives to look out for tuna that bore a stamp guaranteeing that they were free of contamination.[8] Despite such assurances, shoppers refrained from buying tuna and the asking price was halved.[9]

Meanwhile, in the USA, Congressman James Van Zandt (Republican, Pennsylvania) who was a member of the US Joint Committee on Atomic Energy revealed that the hydrogen bomb explosion was equivalent to around 14 million tons of TNT. The actual figure was some 15 million tons which made the bomb 1000 times greater than that dropped on Hiroshima. It was reported that W. Sterling Cole (Republican, New York), who was Chairman of the Committee, suggested that the Lucky Dragon might have been spying.[10] It was apparent to all that there were conflicting statements being made about the voyage of the Lucky Dragon. Amid American claims of negligible danger from consumption of fish caught outside of the test area, physicist Taketani Mituo responded that the US Ambassador to Tokyo, John M. Allison, should be sent some highly contaminated fish and made to eat it.[11] To make matters worse, other fishing boats such as the Kōei-maru arrived at the port of Misaki on 27 March 1954 with radioactive tuna. And Kimura's laboratory at the University of

Tokyo detected radioactive elements in the ash from the Lucky Dragon, including the deadly Strontium-90. In mid-May, the US government announced that the Castle series of bomb tests in the Pacific had come to an end but the harm had been done. On 23 September 1954, the Lucky Dragon's chief radio operator Kuboyama Aikichi passed away. The remaining crewmen were released from hospital on 10 May 1955.[12]

During this time, the Japanese illustrated magazine *Asahi Gurafu* (*Asahi Graph*) ran over ten articles that sought to address the fears of the Japanese public about what had occurred. The articles gave readers almost a televisual experience with black and white photographs and informative captions conveying a sense of being there, witnessing what had been captured by the camera. An article entitled "Dai-san no genbaku-ka" ("The Third Atomic Bomb Disaster") appeared in the 31 March 1954 issue of *Asahi Gurafu*. Photographs spread over four pages conveyed the widespread perception that the Japanese had once again fallen victim to a nuclear weapon.[13] The final page is devoted to the concern about "atomic tuna" and the dangers that it posed to consumers. We see Nishiwaki Yasushi and his American wife Jane Fischer measuring radiation levels of the Lucky Dragon at Yaizu. They had arrived there in the morning of 17 March after Nishiwaki had rushed to the Osaka Central Market the day before and detected that the tuna were radioactive. The next photograph shows a fishmonger's shop with a large sign assuring that the shop does not buy "atomic tuna." The final photograph shows how journalists and cameramen flocked to Yaizu to cover the story, being so bold as to even go on board the fishing boat despite the possible danger.[14] In this way, the article has a documentary-like quality, allowing readers to view key aspects of the story as it developed and as it was covered by the media itself.

A week later, the 7 April 1954 issue of *Asahi Gurafu* focused on how the public were concerned that they had become atomic victims, too, as a result of having eaten contaminated tuna. The article entitled "Genbaku maguro gojitsudan" ("Atomic Tuna Story Sequel") explained how more than half of the tuna that had reached the Osaka Central Market had made its way to fishmongers in Tondabayashi, Osaka prefecture, before the prefectural public health office had contacted the market about the contamination. Over 243 people from Tondabayashi subsequently learned that they had eaten radioactive tuna.[15]

The main photo accompanying the article shows a room full of people, young and old, awaiting medical examination. Their anxious, unsmiling faces signalled to readers that it was the fear of the unknown that was so concerning. Some had spent sleepless nights worrying that their hair might fall out.[16]

In the 19 May 1954 issue of *Asahi Gurafu*, a double-page spread entitled "Shi no hai ga kabutta Nihon gyosen" ("The Japanese Fishing Boats Covered with Ashes of Death") was devoted to providing readers with the details of sixteen deep-sea fishing boats, including the Lucky Dragon, that had been affected by radioactive fallout. The two pages were divided up into a grid-like, rectangular layout: brief details of each vessel were given along with a photograph of the ship in question, photographs of the respective ship's captain, along with his name and age. The message of the article was immediately clear. The Lucky Dragon was not the only fishing boat affected.[17] Thirty-one contaminated fishing boats had returned to Japan during the period from 15 March through to the end of April. They had been detected as part of strict rules regarding the examination of deep-sea fishing boats at five government-designated ports. As a result, 142 tons of radioactive fish had been disposed of. The article reported how the US government would continue hydrogen bomb tests until the end of June 1954.

Writing for the *Bulletin of the Atomic Scientists* the following year, the American scholar Herbert Passin sought to contextualize the Lucky Dragon Incident within the broader US-Japan relationship. He criticized the Japanese for not really wanting to know that American soldiers and Marshall Islanders were also affected by the nuclear fallout at Bikini Atoll. He suggested that most Japanese preferred to think of themselves as unique, privileged victims of American atomic policy. Americans seemed to think that the Japanese were being unnecessarily hysterical about the radioactive tuna.[18]

The American-initiated *Family of Man* exhibition which toured Japan sought to remind the Japanese that they were not unique. Even when it came to nuclear weapons, it was a danger that was posed to all. As we will see, Japanese attempts to highlight what occurred at Hiroshima and Nagasaki would ultimately be thwarted.

The Family of Man

The 1955 exhibition *The Family of Man* which was initially shown at the Museum of Modern Art in New York City served to counter the anti-nuclear fervour generated by the Lucky Dragon Incident. The end of the Allied Occupation signalled that Japan had rejoined the family of nations.

The Family of Man exhibition made this narrative visible to the world through the inclusion of images of the Japanese people alongside those of other peoples. As we shall see, how much of the past could be shown was a matter of some controversy. The beginning of the atomic age could only be hinted at as was the spectre of nuclear war that was facing humankind. Rather, the exhibition was geared more to reminding people of their common humanity.

To understand the controversy surrounding the Japanese iteration of the exhibition, we need to go back to 9 August 1945, shortly after the atomic bomb had been dropped on Nagasaki. The young Japanese military photographer Yamahata Yōsuke rushed there by train to record the devastation. He spent some eight hours taking 121 photographs, some of which were published in the newspaper the *Mainichi Shimbun* on 21 August, but most of which were only made public after the end of the Allied Occupation in 1952 when they were published in *Life* magazine on 9 September 1952 and in book form.[19]

Meanwhile in July 1947, the renowned American photographer Edward Steichen became director of the Photography Department at the Museum of Modern Art in New York City, serving in that role until 1962. The most important exhibition that he undertook during that time was the exhibition *The Family of Man*, an expanded version of which travelled to Japan. Funding for the initial exhibition was originally sought as early as 1950 and the working title of the exhibition was *Image of America*.[20] Steichen had curated several war-related exhibitions and he was determined to present the war in a more positive light that embraced life and emphasized how alike people were throughout the world.

Drawing on the term "family of man" from a speech given by Abraham Lincoln that he found in Carl Sandburg's biography, Steichen surveyed and solicited images from throughout the world,[21] choosing 503 photographs from 68 countries, including Yamahata's image of a forlorn child in Nagasaki holding a rice ball which seemed to have relatively little overt political content. In order to offset cold war fears and cultural differences which had been emphasized during World War II, Steichen chose to embrace universalism and engender a sense of hope by emphasizing the American values of family unity and democracy.[22] The exhibition opened at MoMA on 26 January 1955 and continued through to 8 May 1955. The exhibition toured the USA to great acclaim.

In the exhibition, Yamahata's portrait of the boy was one of nine close-up portraits of the same size hung together at MoMA, with a mirror at the centre. These images captured the faces of people (three women, three children and three men) suffering in various ways. Visitors could see their own faces reflected among them. A caption in the distance quoting the words of Sophocles read: "Who is the slayer, who the victim? Speak." This prompted viewers to question and reflect on their own position.[23] The mirror was removed after two weeks. Alongside the portraits were the words of Bertrand Russell that warned of how war with hydrogen bombs would most likely bring about the end of the human race. In the book, too, Russell's words appear alongside the portraits.[24]

Towards the end of the exhibition at MoMA, visitors entered a small, dark gallery where they encountered a 6 × 8 foot, dramatically back-lit transparency of a 1952 hydrogen bomb test at Enewetak Atoll that had first been published in *Life* magazine in 1954.[25] As the only colour image in the entire exhibition, it was difficult to miss. Located close to the exit, it served as a climax to the exhibition.[26] A Japanese visitor to New York, Hirano Takeshi reported his impressions in *Kamera Handobukku* (*Camera Handbook*) published in October 1955. He mentioned that the people he came across at the exit seemed to have the appearance of having shared a special, intimate moment. They left the exhibition hushed but deeply moved. They seemed to have come to an understanding about the future of humankind.[27] In contrast, the American art critic Hilton Kramer criticized the exhibition for being

> a self-congratulatory means for obscuring the urgency of real problems under a blanket of ideology which takes for granted the essential goodness, innocence, and moral superiority of the international "little man,"... the abstract disembodied hero of a world-view which regards itself as superior to mere politics.[28]

Kramer ruminated whether the exhibition marked a new kind of politics in the air. The seemingly apolitical exhibition was in fact highly political. Kramer saw the exhibition as "a reassertion in visual terms of all that has been discredited in progressive ideology."[29] What's more, he didn't see the exhibition as necessarily contributing to a better understanding of the art of photography and its aesthetic status. Rather, it seemed to add to the confusion about the difference between photography as art and photography as journalism.

The exhibition coincided with a period in the mid-1950s when the US Information Agency (USIA), which had been established in August 1953, began to play a greater role in American foreign policy. The exhibition provided the opportunity for an international cultural initiative which while showing the unity of humankind would nevertheless portray the USA in a largely positive light as a model for the rest of the world. A benevolent image of the nation and the benefits of capitalism were conveyed in press and film, as well as exhibitions and trade fairs. Indeed, W.T. Lhamon likens the exhibition to a promotional trade fair![30]

From New York City to Tokyo

Several versions of the exhibition were sent abroad by the USIA. To facilitate this, the MoMA version of the exhibition was sold to the USIA and the USIA specially commissioned versions of the exhibition[31] which also travelled to Japan in 1956 where the USIA (known in Japan and outside of the USA as the USIS) also toured the *Atoms for Peace* exhibit that same year. Despite the universal theme behind the exhibition and perception that duplicates were toured, the USIA did adapt the exhibition when it came to tour Japan. This was in line with its country-by-country information strategy. There was recognition that most of the images in the original exhibition had been taken by American or European photographers and 200 of the 503 images were taken in the USA. MoMA provided negatives and layout to the Japanese organizers who in turn asked G.T. Sun, the commercial studio run by Yamahata Yōsuke and his father Shōgyoku,[32] to prepare the prints for a year-long exhibition sponsored by the newspaper *Nihon Keizai Shimbun* as part of its eightieth anniversary celebrations.

As part of preparations to tour the exhibition in Japan, Steichen arrived in Japan on 7 August 1955 to meet with local organizers of the exhibition and representatives of the Japanese photography world. During his visit, Steichen expressed an interest in organizing an exhibition of Japanese photography at MoMA as well as adding the work of Japanese photographers to *The Family of Man* exhibition that would tour Japan. There had been about ten images of Japan in the MoMA exhibition but little in the way of work by Japanese photographers apart from Yamahata Yōsuke's portrait of a boy at Nagasaki.[33] Japanese photography magazines and photobooks were sent to his room at the Imperial Hotel in Tokyo where he was staying. On his return to the USA, Steichen wrote to the editor of *Nihon Keizai Shimbun*, Enjōji Jirō, that he wanted to add a total of fifty-five photographs to the exhibition.[34] The extra images by Japanese photographers helped to remedy the lack of images of Japan by the Japanese and made for a total of 560 photographs in the exhibition. This served to heighten Japanese interest in the exhibition.

Sales of Steichen's 1955 book in Japan had already encouraged interest even before the exhibition had opened in Tokyo. The book was sold in Japan with a commentary in Japanese inserted that included translations of the quotes. An initial 20,000 copies went on sale on 15 January 1956 but sold out in three days, reaching No. 3 on the best sellers list. A further 20,000 copies were made available on 27 February but also sold out.[35] On 21 March, the first day that the exhibition was open to the public, 300 people waited outside the Takashimaya Department Store for the doors to open at 10:00 a.m. at which time they raced to the elevator to reach the exhibition on the 8th floor. Over 2000 people were admitted to the exhibition in just two hours.[36] Fifteen hundred copies of the book that had been set aside for the exhibition were sold out that morning, so most of the visitors who attended the exhibition bought the book. In the evening of 22 March, a further consignment of books arrived at the port of Yokohama for sale from 25 March. Not only was the exhibition a great success but Steichen's book was a publishing phenomenon.

Among the 2000 people who viewed the exhibition that morning, there were 630 girls and boys from Mita Senior High School who were attending as part of their Social Studies. One of their teachers explained that two months previously, the school principal had praised the exhibition in front of the students and emphasized the captions that accompanied the photographs which he no doubt had seen in the book. A photograph showing the head-tops of many of the students viewing the exhibition was carried in the evening edition of *Nihon Keizai Shimbun*.[37]

Many of them would have purchased the small Japanese exhibition catalogue produced by *Nihon Keizai Shimbun*. In the form of a small square booklet, it only offered a very limited selection of images. They did however include Carl Mydan's photograph of a farming family in Japan which was juxtaposed next to Nat Farbman's photograph of an extended family in Bechuanaland (later known as the Republic of Botswana). Both images had also appeared in Steichen's original 1955 book in the same order but on consecutive pages which gave them a different effect. In the Japanese publication, the Japanese rural family was compared with a family from Sicily, Italy, showing Japan as part of the developed world rather than suggesting parallels with a scantily clad family from underdeveloped Africa.[38] Donald Richie, who reviewed the exhibition for *Nippon Times*, reported that the family photographs, along with that of an American family, were arranged in a square, and visitors would move around the space to view them. He argued that "the 'oneness' of mankind' at once becomes evident" despite each group wearing different kinds of clothes.[39]

As for the work of Japanese photographers in the Japanese catalogue, only Yamahata Yōsuke's image of the blood-flecked face of a young boy at Nagasaki was included. However, rather than appearing as the centrepiece of a mosaic of nine people against a white page as in the American catalogue,[40] we see a cropped version of the photograph highlighting the child's pale face against a black background in the Japanese version. It is the last image that appears in the catalogue and a memorable one at that.[41]

In the exhibition in Japan, a large black and white image of an exploding atomic bomb that had been shown at other venues in place of the colour hydrogen bomb transparency was in turn replaced by a large photographic mural of images of the devastation at Nagasaki taken by Yamahata.[42] This suggested a change of perspective from victor to victim and the recognition that images represent a particular point of view. This was done by the Japanese organizers with the agreement of Steichen who had left it to their discretion.[43] As the architectural photographer Watanabe Yoshio who was a member of the organizing committee wrote in a special issue of the magazine *Sankei kamera* devoted to the exhibition, architect and fellow committee member Tange Kenzō had taken on the task of how to include photographs depicting the effects of the atomic bomb that Steichen had been strongly moved by when he visited Japan. Watanabe and other photographer members of the committee had only become aware of Tange's solution when the exhibition was being installed.

The images were curtained off by a white cloth when the Emperor Hirohito viewed the exhibition in the morning of 23 March 1956 at the Takashimaya Department Store in Nihonbashi, Tokyo. The cloth served to protect the imperial gaze from what was deemed to be a disturbing sight. The emperor was accompanied by the president of the *Nihon Keizai Shimbun* newspaper company, Yorozu Naoji, John M. Allison, the US Ambassador to Japan, and the Director of the National Museum of Modern Art, Okabe Nagakage. The article made no mention of the curtain and censored images but did quote the emperor as saying that he hoped that the exhibition would contribute to deepen US-Japan relations and contribute to world peace. Above the article was a photograph of the emperor looking upwards at photographs that were displayed within a suspended, circular panel within the exhibition as if to note the lofty and transcendent theme of the exhibition.[44]

Although the exhibition sponsor *Nihon Keizai Shimbun* appears to have been largely silent in terms of coverage of the controversy, the evening edition of rival newspaper the *Asahi Shimbun* noted that the special exhibit of five photographs showing the destruction at Nagasaki had been curtained off at the instigation of the Japanese organizers despite having been on view from the opening day. However, there had apparently been various criticisms, the matter was reconsidered, and the decision was made not to show the images to the emperor. The curtain was removed once the emperor left. According to the Imperial Household, the emperor had seen the exhibition catalogue prior to his visit and was expecting to see such images. However, they had heard from Takashimaya Department Store that morning about the curtain but it was something that those sponsoring the exhibition, *Nihon Keizai Shimbun*, instigated on their own without their prompting. The head of the newspaper's Planning Department tried to explain it away by suggesting that the exhibition display had only been finalized that day, prior to which it was still being fine-tuned despite having been open to the public for two days. The article also mentioned that there had been criticisms that the effect of the Nagasaki images was too weak and too strong. Given that the matter was still being considered, the decision was made not to show the images to the emperor. Ironically, the newspaper article was accompanied by a photograph showing two schoolgirls viewing four of those censored images.[45]

Fred Saito, writing in *Nippon Times*, suggested that the curtaining of the images was reminiscent of pre-World War II days when palace chamberlains removed all disturbing stories from newspapers before they were sent on to the emperor. A spokesman for *Nihon Keizai Shimbun* denied that use of the white cloth constituted censorship. Rather, it was a decision made after two days of "furore." The newspaper acknowledged that there was the view (actually expressed by Prince Mikasa, the youngest brother of the emperor) that the choice of atomic-bomb photos was "weak" in terms of conveying the destruction of what had occurred at Nagasaki.[46] This was in reference to Prince Mikasa's comments reported in the *Nihon Keizai Shimbun* on 21 March, after he had officially cut the tape and opened the exhibition the afternoon before. Prince Mikasa noted the photo mural which included Yamahata's image of the young boy but thought that a mushroom cloud and a-bomb victims with keloid scars would have been more effective.[47] Alternatively, others deemed the images as being too strong and too much of a departure from the theme of the exhibition.[48] Others were drawn to other aspects of the exhibition. Among the dignitaries at the opening was the Chief Justice of the Supreme Court, Tanaka Kōtarō, who likened the experience of the exhibition to that of seeing a movie. He commented on how the theme that ran through the exhibition had a universal quality that resonated with his own beliefs.[49] All in all, the reception of the exhibition was generally a positive one, although not without controversy.

Soon after the curtain incident, Steichen decided to permanently remove the confronting atomic-bomb images from the exhibition.[50] Yamahata's portrait of the little Nagasaki boy holding a rice ball that had been part of the original MoMA exhibition was shown by itself rather than as part of a wall of faces as seen at MoMA. The boy was no longer part of a larger photograph with what may have been his stricken mother which had been an element of the photo-mural. Now removed out of context and hanging on its own as a close-up portrait, the forlorn boy was flanked on either side by blank velvet walls.

Tange Kenzō who had been responsible for adapting the original exhibition layout for the Tokyo venue, as well as designing the portable display system which could be removed and used elsewhere, said to the *Asahi Shimbun* that they had obtained Steichen's approval the previous year for including the photographs but in light of Steichen's recently expressed wishes, the organizers had decided to remove them. Tange thought, nevertheless, that the offending photographs did not run counter to the theme of the exhibition, nor did they interfere artistically with the overall feel of the exhibition.[51] Donald Richie suggests that this made the photograph even more powerful and much more forbidding than would have been the case if the wall had been filled with the "most shocking of pictures."[52] So in a way, the single image of the boy as seen in the Japanese catalogue, and as ultimately appeared in the exhibition at Tokyo, speaks to the destructive force of the atomic bomb, the absence of the other photographs and the act of censorship that occurred.

The Family of Man exhibition would be the last major exhibition that the USIS chief exhibits officer based at the US Embassy in Tokyo, Frances Blakemore, would work on. She was, apart from Steichen himself, the only non-Japanese member of the small organizing committee.[53] Watanabe Yoshio who was a fellow committee member revealed in his article for *Sankei kamera* that US Secretary of State John Foster Dulles had attended the special viewing to celebrate the opening of the exhibition on 20 March 1956.

Dulles had met with the Japanese Foreign Minister Shigemitsu Mamoru and nine other top Japanese officials for a two and a half hour meeting a couple of days previously on 18 March in Tokyo. He had come from Seoul and was on his way back to Washington, DC, after a nine-nation tour of Asia that included attending the Southeast Asian Treaty Organization (SEATO) conference in Karachi, Pakistan.[54] At the meeting, Dulles was accompanied by nine other Americans including Andrew Berding, Assistant Director of the United States Information Agency (USIA) and Ambassador Allison. Topics included changes in Soviet policy in Southeast Asia, his impressions of Taiwan and the Republic of Korea, Communist China's ambitions in the region and Japan-Soviet negotiations. The meeting concluded with Dulles saying that it had been informative and useful, more so "than any talk he had had in Japan since his work on the Japan Peace Treaty in 1951."[55] Despite Dulles' good impressions on leaving the meeting, there was a sour note.

At the exhibition, Dulles noted that it differed from the version that had been shown in the USA and Europe. He asked whether the photographs of the effects of the atomic bomb had been included at the direction of Steichen and Frances Blakemore who also was in attendance wasn't able to offer much of a response. Watanabe suggests that she had admired the exhibition but nevertheless, a request was made by the Americans to the exhibition sponsor *Nihon Keizai Shimbun* that the photographs not be shown on 23 March when Ambassador Allison accompanied Emperor Hirohito for a viewing of the exhibition. Whether Blakemore personally experienced any flak at the US Embassy as a result of the controversy is not known.[56]

In Watanabe's opinion, if the curtain hiding the photo-mural had been left that way until Steichen had indicated what he wanted to be done, there would not have been as much controversy. As it turned out, there was the appearance that the images were only hidden from the gaze of the emperor. As Watanabe's article "Tennō heika ni kankei wa nai"[57] made plain, it had nothing to do with the emperor. Nevertheless, the evening edition newspapers made it a controversy, and news of it was conveyed to Steichen in the USA who was very surprised and decided to have the photographs removed. This was despite long-distance phone calls made by the organizers to explain the situation in Japan and to continue to exhibit them. During this time, the organizing committee was not consulted.[58]

Watanabe noted that Steichen had acknowledged that there were other war-related photographs in the exhibition but they tended to be aesthetically symbolic images. In comparison, the interest of the Nagasaki photographs rested too much in their realism. That expression was not in harmony with the rest of the exhibition. When Watanabe and other photographers on the organizing committee viewed the exhibition, he conceded that they also felt that.[59]

Meanwhile, the *Atoms for Peace* exhibit that Blakemore helped design was on its two-year tour of Japan. She would retire the following year after having facilitated these two very successful exhibitions.[60] Despite the controversy, crowds continued to flock to the exhibition. The queue of people waiting to gain admission stretched from the 8th floor to the rooftop of the department store building.[61] But rival newspaper *Asahi Shimbun* continued to report about the censorship of the photographs. It reported Steichen as saying from New York that when he had visited Japan last autumn, he had been deeply moved by the photographs. However, in terms of the exhibition, he wanted to avoid drawing attention to a specific event which might detract from the overall theme.[62]

Tokyo and Beyond

Ultimately, the atomic bomb photo mural was removed, and four versions of *The Family of Man* exhibition (with images in different sizes) toured Japan over the period of one year. Large versions of the exhibition were shown in Tokyo, Osaka, Nagoya, Fukuoka, Kyoto, Okayama, Hiroshima and Shizuoka in 1956. Venues included the Takashimaya Department Store in Osaka (8–20 May), the Matsuzakaya Department Store in Nagoya (29 May to 10 June), Takashimaya Department Store in Tokyo for a second time (17–29 July). When the exhibition was shown in Hiroshima, some 28,700 people viewed the exhibition over two weeks in October.[63] Smaller versions of the exhibition were shown through to 1957 in more regional locations: Sendai, Akita, Hakodate, Niigata, Sapporo, Kokura, Kagoshima, Miyazaki, Kumamoto, Ōita, Yokohama, Utsunomiya, Matsuyama, Takamatsu, Ehime, Himeji, Tottori and Kanazawa.[64] Steichen estimated that more than 1 million people saw the exhibition in Japan.[65]

Steichen famously suggested that those who looked at the pictures in the exhibition were involved in an act of audience participation—the audience viewed the photographs and the people in the photographs looked back at the audience. Both sides recognized each other. He attributed that insight to a Japanese poet who said that "when you look into a mirror, you do not see your reflection, your reflection sees you."[66] While these words might seem cryptic, they make sense in light of Japanese commentary on the exhibition. The photographer Natori Yōnosuke suggested that the exhibition reflected a certain viewpoint. Rather than the images necessarily speaking for themselves, the selection of images reflected how Americans saw other peoples of the world. Rather than necessarily seeing themselves, some Japanese saw Americans reflected back.[67]

Conclusion

The Family of Man was a great success for both Steichen and the USIS/USIA. By 1960, Steichen estimated that some nine editions of the exhibition had been seen by 7 million people in twenty-eight countries. It ultimately toured the world until the early 1960s and attracted over 9 million visitors.[68] While people were only starting to accept photography as art, Steichen saw the exhibition's success as demonstrating its importance as a visual means of mass communication without peer.[69] On this point he seemed to be in agreement with Hilton Kramer. It would be wrong, however, to simply assume that the exhibition would be popular in Japan if it was successful elsewhere. The inclusion and then later exclusion of Yamahata's photographs of Nagasaki from the exhibition show that Steichen and the local organizers gave careful thought to what Japanese audiences would be receptive to. The exclusion of Yamahata's photographs from exhibitions in Japan demonstrates the sensitivities regarding the depiction of the effect of the atomic bombs on humans. This has resulted in a type of national amnesia and lack of specificity in Japan and the USA about what happened at Hiroshima and Nagasaki and the possible risks associated with nuclear power. The Lucky Dragon Incident was crucial in reminding the Japanese people soon after the end of the Allied Occupation what had occurred in Japan and at Bikini Atoll.

Notes

1. Kyodo-UP, "H-Bomb Test Observers to Leave Next Weekend," *Nippon Times*, 1 Mar. 1954, 1.
2. Ralph E. Lapp, with a foreword by Pearl S. Buck, *The Voyage of the Lucky Dragon* (Harmondsworth, Middlesex: Penguin Books, 1958), 6.
3. Lapp, *The Voyage of the Lucky Dragon*, 34–35. See also Dwight Martin, "First Casualties of the H-Bomb," *Life* 36, no. 13 (29 Mar. 1954): 17, 19–21.

4. Lapp, *The Voyage of the Lucky Dragon*, 34–35, 75–76.
5. "U.S., Japan Probing Bikini Bomb Incident," *Nippon Times*, 17 Mar. 1954, 1–2.
6. "U.S., Japan Probing Bikini Bomb Incident."
7. Lapp, *The Voyage of the Lucky Dragon*, 85, 96.
8. "OK Sounded On All Fish," *Nippon Times*, 19 Mar. 1954, 1.
9. "Wait-See Policy Set on Bomb Row," *Nippon Times*, 19 Mar. 1954, 1.
10. Lapp, *The Voyage of the Lucky Dragon*, 105, 109.
11. Lapp, *The Voyage of the Lucky Dragon*, 113, 116.
12. Lapp, *The Voyage of the Lucky Dragon*, 116, 118, 151.
13. "Dai-san no genbaku-ka" ("The Third Atomic Bomb Disaster"), *Asahi Gurafu* (*Asahi Graph*), no. 1545 (31 Mar. 1954): 4–7, esp. 4.
14. "Dai-san no genbaku-ka," 7.
15. "Genbaku maguro gojitsudan" ("Atomic Tuna Story Sequel"), *Asahi Gurafu* (*Asahi Graph*), no. 1546 (7 Apr. 1954): 6–7.
16. "Genbaku maguro gojitsudan."
17. "Shi no hai ga kabutta Nihon gyosen" ("The Japanese Fishing Boats Covered with Ashes of Death"), *Asahi Gurafu* (*Asahi Graph*), no. 1552 (19 May 1954): 6–7; Matashichi Ōishi, trans. by Richard H. Minear, *The Day the Sun Rose in the West* (Honolulu: University of Hawai'i Press, 2011), 30.
18. Herbert Passin, "Japan and the H-Bomb," *Bulletin of the Atomic Scientists* 11, no. 8 (Oct. 1955): 289–292.
19. Ayelet Zohar, "The Day After," accessed 12 Sept. 2017, http://www.endofempire.asia/0809-2-the-day-after-4/
20. Fred Turner, "*The Family of Man* and the Politics of Attention in Cold War America," *Public Culture* 24, no. 1 (2012): 55–84, esp. 74–75.
21. Edward Steichen, *A Life in Photography* (London: W.H. Allen, in collaboration with the Museum of Modern Art, 1963), Chapter 13, three pages following plate 225.
22. John H. Herz, *International Politics in the Atomic Age* (New York: Columbia University Press, 1959), 309.
23. Eric J. Sandeen, "*The Family of Man* at the Museum of Modern Art: The Power of the Image in 1950s America," *Prospects* 11 (Oct. 1986): 367–391, esp. 372; for a view of the installation see Mary Anne Staniszewski, *The Power of Display: A History of Exhibition Installations at the Museum of Modern Art* (Cambridge, Mass.: MIT Press, 1998), 246.
24. Reflecting the exhibition layout, the words of Sophocles appear on the page after the quote from Bertrand Russell. See Edward Steichen, *The Family of Man: The Greatest Photographic Exhibition of All Time, 503*

Pictures from 68 Countries, Created by Edward Steichen for the Museum of Modern Art (New York: Maco Magazine Corp. for the Museum of Modern Art, 1955), 180.

25. John O'Brian, "The Nuclear Family of Man," *The Asia-Pacific Journal: Japan Focus* 6, no. 7 (2 Jul. 2008), accessed 15 Sept. 2017, http://apjjf.org/-John-O'Brian/2816/article.html
26. Turner, "*The Family of Man*," 80; Lili Corbus Bezner, *Photography and Politics in America: From the New Deal into the Cold War* (Baltimore: Johns Hopkins University Press, 1999), 154–155.
27. Hirano Takeshi, "Sutaiken no 'Za famirii obu man' shashin-ten no inshō" ("Impressions of the Photographic Exhibition 'The Family of Man'", *Kamera handobukku*, special issue no. 23 (Oct. 1955), "Kyō to shashin" ("Photography and Today"): 65–69, esp. 69.
28. Hilton Kramer, "On the Horizon: Exhibiting the Family of Man," *Commentary* 20, no. 4 (Oct. 1955): 364–367, esp. 367.
29. Kramer, "On the Horizon: Exhibiting the Family of Man," 367.
30. W.T. Lhamon, Jr., *Deliberate Speed: The Origins of a Cultural Style in the American 1950s* (Washington, D.C.: Smithsonian Institution Press, 1990), 147.
31. Museum of Modern Art, New York, International Council and International Program Records Subseries I.B: International Program SP-ICE Exhibition Files, through Series VI: International Council Administrative Records, SP-ICE-10-55 description, accessed 12 Sept. 2017, https://www.moma.org/learn/resources/archives/EAD/ICIP_SeriesIB_VIf
32. Mark Silver, "Framing the Ruins: The Documentary Photographs of Yamahata Yōsuke (Nagasaki, August 10, 1945)," in *Imag(in)ing the War in Japan: Representing and Responding to Trauma in Postwar Literature and Film*, eds. David Stahl and Mark Williams (Leiden: Brill, 2010), 229–268, esp. 237.
33. Bezner, *Photography and Politics in America*, 142.
34. Enjōji Jirō, "'Za famirii obu man' ni yoseru" ("The Forthcoming 'Family of Man' Exhibition"), *Bungei Shunju* 34, no. 2 (Feb. 1956): 182–183.
35. "Shashinshū tachimachi urekire" ("Photobook Immediately Sells Out"), *Nihon Keizai Shimbun*, 21 Mar. 1956, evening edition, 1.
36. "Kaimaku nijikan de ni sen mei" ("2000 People Admitted in the First Two Hours"), *Nihon Keizai Shimbun*, 21 Mar. 1956, evening edition, 1.
37. "Sōchō kara gotta kaeshi: Za famirii obu man shashin ten" ("In Return for the Early Morning Chaos: The Family of Man Photographic Exhibition"), *Nihon Keizai Shimbun*, 22 Mar. 1956, evening edition, 3.
38. Steichen, *The Family of Man* (1955), 57–58; also Edward Steichen, *The Family of Man: The 30th Anniversary Edition of the Classic Book of*

Photography Created by Edward Steichen for the Museum of Modern Art, New York (New York: The Museum of Modern Art, 1986), 57–58.

39. Donald Richie, "'Family of Man' Exhibition: Living Photographic Essay," *Nippon Times*, 9 Apr. 1956, 4.
40. Steichen, *The Family of Man* (1955, 1986), 178.
41. Edward Steichen, *Za famirii obu man: Warera mina ningen kazoku* (*The Family of Man: We're All Part of the Human Family*) (Tokyo: Nihon Keizai Shimbun-sha Kikaku-bu, 1956).
42. Allan Sekula, "The Traffic in Photographs," *Art Journal* 41, no. 1 (Spring 1981): 15–25, esp. 19.
43. This is revealed by Kimura Ihei, a member of the local organizing committee in "Omunibasu zadankai: Saikin no wadai of kataru" ("Omnibus Round-Table Discussion: Recent Topics"), *Asahi kamera* 41, no. 6 (June 1956): 124–131, esp. 127.
44. "Tennō heika gokanshō: Za famirii obu man ten" ("The Emperor Enjoys: The Family of Man Exhibition"), *Nihon Keizai Shimbun*, 23 Mar. 1956, evening edition, 4.
45. "Tennō heika ni wa misenai: Genbaku shashin ni kaaten" ("Not Shown to His Majesty the Emperor: A-Bomb Photos are Curtained Off"), *Asahi Shimbun*, 23 Mar. 1956, evening edition, 7.
46. Fred Saito, "At 'Family of Man' Show: Pictures of A-Victims Hidden from Emperor," *Nippon Times*, 24 Mar. 1956, 3. For a photograph of the emperor accompanied by Yorozu and Allison, see "To Promote Friendship," *Nippon Times*, 24 Mar. 1956, 1.
47. Prince Mikasa, "Ari no mama no ningen no sugata" ("The Stark Human Condition"), *Nihon Keizai Shimbun*, 21 Mar. 1956, 11.
48. Saito, "At 'Family of Man' Show." For a photograph of the emperor accompanied by Yorozu and Allison, see "To Promote Friendship."
49. Tanaka Kōtarō, "Eiga no yō da," *Nihon Keizai Shimbun*, 21 Mar. 1956, 11.
50. Tessa Morris-Suzuki, *The Past Within Us: Media, Memory, History* (London: Verso, 2005), 115–116.
51. "'Genbaku shashin wa hazuse': Za famirii obu man ten" ("'Remove the A-bomb Photographs': The Family of Man Exhibition"), *Asahi Shimbun*, 27 Mar. 1956, 11.
52. Richie, "'Family of Man' Exhibition", 4. For a USIA photograph of the Tokyo installation, see Eric J. Sandeen, "The International Reception of The Family of Man," *History of Photography* 29, no. 4 (Winter 2005): 344–355, esp. 347.
53. See the catalogue insert, *Za Famirii Obu Man* (*Ningen kazoku*) *Nihongo kaisetsusho* (*The Family of Man Japanese Language Reference Booklet*) (Tokyo: Niyon Keizai Shimbunsha, 1956). The organizing committee members are listed.

54. Sid White, “Dulles Optimistic Over Peace Outlook; Tokyo Talks Wind Up His Tour of East,” *Washington Post*, 19 Mar. 1956, 6.
55. US Government, Department of State, “Memorandum of a Conversation, Tokyo, March 18, 1956,” in *Foreign Relations of the United States, 1955–1957, Japan*, vol. 23, part 1, ed. David W. Mabon (Washington DC: US Government Printing Office, 1991), document 70, 156–163, accessed 8 Aug. 2018, https://history.state.gov/historicaldocuments/frus1955-57v23p1/d70
56. Watanabe, Yoshio. “Tennō heika ni kankei wa nai” (“It’s nothing to do with the Emperor”), *Sankei kamera*, “Family of Man” special issue (June 1956), 137.
57. Watanabe, “Tennō heika ni kankei wa nai.”
58. Watanabe, “Tennō heika ni kankei wa nai..”
59. Watanabe, “Tennō heika ni kankei wa nai.”
60. Michiyo Morioka, *An American Artist in Tokyo: Frances Blakemore, 1906–1997* (Seattle: Blakemore Foundation, 2007), 128.
61. “Tsui ni manin fuda dome: Za famirii obu man ten” (“In the End, Full House: The Family of Man Exhibition”), *Nihon Keizai Shimbun*, 25 Mar. 1956, 11.
62. “‘Tokutei jiken saketai’ Sutaiken-shi no hanashi: Genbaku shashin jogai de” (“Steichen Says He Wishes to Avoid Specific Events: Exclusion of A-Bomb Photographs”), *Asahi Shimbun*, 27 Mar. 1956, evening edition, 3.
63. Eric J. Sandeen, *Picturing an Exhibition: The Family of Man and 1950s America* (Albuquerque: University of New Mexico Press, 1995), 96.
64. Inubuse Masakazu, “‘Ningen Kazoku (Famirii Obu Man)’ ten no Nihon ni okeru juyō” (“On the Reception of the ‘Family of Man’ Exhibition in Japan”), *Ōsaka Geijutsu Daigaku kiyō* (*Proceedings of the Osaka University of the Arts*) 37 (Dec. 2014): 17–28, esp. 18.
65. Steichen, *A Life in Photography*, Chapter 13, three and four pages following plate 225.
66. Steichen, *A Life in Photography*, Chapter 13, five pages following plate 225.
67. “Omunibasu zadankai: Saikin no wadai of kataru” (“Omnibus Round-Table Discussion: Recent Topics”), *Asahi kamera* 41, no. 6 (June 1956): 124–131, esp. 124.
68. Museum of Modern Art, New York, “The Family of Man,” accessed 12 Sept. 2017, https://www.moma.org/calendar/exhibitions/2429?locale=en
69. Edward Steichen, “On Photography,” *Daedalus* 89, no. 1 (Winter 1960): 136–137.

Bibliography

Asahi Shimbun, 23–27 Mar. 1956.

Bezner, Lili Corbus. *Photography and Politics in America: From the New Deal into the Cold War*. Baltimore: Johns Hopkins University Press, 1999.

"Dai-san no genbaku-ka" ("The Third Atomic Bomb Disaster"), *Asahi Gurafu* (*Asahi Graph*), no. 1545 (31 March, 1954): 4–7.

Enjōji, Jirō. "`Za famirii obu man' ni yoseru" ("The Forthcoming `Family of Man' Exhibition"). *Bungei Shunju* 34, no. 2 (Feb. 1956): 182–183.

"Genbaku maguro gojitsudan" ("Atomic Tuna Story Sequel"), *Asahi Gurafu* (*Asahi Graph*), no. 1546 (7 Apr. 1954): 6–7.

Herz, John H. *International Politics in the Atomic Age*. New York: Columbia University Press, 1959.

Hirano, Takeshi. "Sutaiken no 'Za famirii obu man' shashin-ten no inshō" ("Impressions of the Photographic Exhibition `The Family of Man'." *Kamera handobukku*, special issue no. 23 (Oct. 1955), "Kyō to shashin" ("Photography and Today"): 65–69.

Inubuse, Masakazu. "`Ningen Kazoku (Famirii Obu Man)' ten no Nihon ni okeru juyō" ("On the Reception of the `Family of Man' Exhibition in Japan"). *Ōsaka Geijutsu Daigaku kiyō* (*Proceedings of the Osaka University of the Arts*) 37 (Dec. 2014): 17–28.

Kramer, Hilton. "On the Horizon: Exhibiting the Family of Man." *Commentary* 20, no. 4 (Oct. 1955): 364–367.

Lapp, Ralph E. *The Voyage of the Lucky Dragon*. Foreword by Pearl S. Buck. Harmondsworth, Middlesex: Penguin Books, 1958.

Lhamon, Jr., W.T. *Deliberate Speed: The Origins of a Cultural Style in the American 1950s*. Washington, D.C.: Smithsonian Institution Press, 1990.

Martin, Dwight. "First Casualties of the H-Bomb." *Life* 36, no. 13 (29 Mar. 1954): 17, 19–21.

Morioka, Michiyo. *An American Artist in Tokyo: Frances Blakemore, 1906–1997* (Seattle: Blakemore Foundation, 2007), 128.

Morris-Suzuki, Tessa. *The Past Within Us: Media, Memory, History*. London: Verso, 2005.

Museum of Modern Art, New York. "The Family of Man." Accessed 12 Sept., 2017. https://www.moma.org/calendar/exhibitions/2429?locale=en.

Museum of Modern Art, New York. International Council and International Program Records Subseries I.B: International Program SP-ICE Exhibition Files.

Nihon Keizai Shimbun, 21–25 Mar. 1956.

Nippon Times, 1–19 Mar. 1954, 24 Mar. 1956, 9 Apr. 1956.

O'Brian, John. "The Nuclear Family of Man." *The Asia-Pacific Journal: Japan Focus* 6, no. 7 (2 July, 2008). Accessed 15 Sept., 2017 http://apjjf.org/-John-O'Brian/2816/article.html.

Ōishi, Matashichi. *The Day the Sun Rose in the West.* Translated by Richard H. Minear. Honolulu: University of Hawai'i Press, 2011.

"Omunibasu zadankai: Saikin no wadai of kataru" ("Omnibus Round-Table Discussion: Recent Topics"). *Asahi kamera* 41, no. 6 (June 1956): 124–131.

Passin, Herbert. "Japan and the H-Bomb." *Bulletin of the Atomic Scientists* 11, no. 8 (Oct. 1955): 289–292.

Sandeen, Eric J. *Picturing an Exhibition: The Family of Man and 1950s America.* Albuquerque: University of New Mexico Press, 1995.

Sandeen, Eric J. "*The Family of Man* at the Museum of Modern Art: The Power of the Image in 1950s America," *Prospects* 11 (Oct. 1986): 367–391

Sandeen, Eric J. "The International Reception of the Family of Man." *History of Photography* 29, no. 4 (Winter 2005): 344–355.

Sekula, Allan. "The Traffic in Photographs." *Art Journal* 41, no. 1 (Spring 1981): 15–25.

"Shi no hai ga kabutta Nihon gyosen" ("The Japanese Fishing Boats Covered with Ashes of Death"), *Asahi Gurafu* (*Asahi Graph*), no. 1552 (19 May, 1954): 6–7

Silver, Mark. "Framing the Ruins: The Documentary Photographs of Yamahata Yōsuke (Nagasaki, August 10, 1945)." In *Imag(in)ing the War in Japan: Representing and Responding to Trauma in Postwar Literature and Film.* Edited by David Stahl and Mark Williams, 229–268. Leiden: Brill, 2010.

Staniszewski, Mary Anne. *The Power of Display: A History of Exhibition Installations at the Museum of Modern Art.* Cambridge, Mass.: MIT Press, 1998.

Steichen, Edward. *A Life in Photography.* London: W.H. Allen, in collaboration with the Museum of Modern Art, 1963.

Steichen, Edward. "On Photography." *Daedalus* 89, no. 1 (Winter 1960): 136–137.

Steichen, Edward. *The Family of Man: The Greatest Photographic Exhibition of All Time, 503 Pictures from 68 Countries, Created by Edward Steichen for the Museum of Modern Art* (New York: Maco Magazine Corp. for the Museum of Modern Art, 1955.

Steichen, Edward, *The Family of Man: The 30th Anniversary Edition of the Classic Book of Photography Created by Edward Steichen for the Museum of Modern Art, New York.* New York: The Museum of Modern Art, 1986.

Steichen, Edward. *Za famirii obu man: Warera mina ningen kazoku* (*The Family of Man: We're All Part of the Human Family*). Tokyo: Nihon Keizai Shimbun-sha Kikaku-bu, 1956.

Turner, Fred. "*The Family of Man* and the Politics of Attention in Cold War America." *Public Culture* 24, no. 1 (2012): 55–84.

US Government, Department of State, "Memorandum of a Conversation, Tokyo, March 18, 1956." In *Foreign Relations of the United States, 1955–1957, Japan,* vol. 23, part 1. Edited by David W. Mabon, Document 70, 156–163. Washington

DC: US Government Printing Office, 1991. Accessed 8 Aug. 2018. https://history.state.gov/historicaldocuments/frus1955-57v23p1/d70.

Washington Post, 19 Mar. 1956.

Watanabe, Yoshio. "Tennō heika ni kankei wa nai" ("It's nothing to do with the Emperor"), *Sankei kamera*, "Family of Man" special issue (June 1956), 137.

Za Famirii Obu Man (*Ningen kazoku*) *Nihongo kaisetsusho* (*The Family of Man Japanese Language Reference Booklet*). Tokyo: Niyon Keizai Shimbunsha, 1956.

Zohar, Ayelet. "The Day After." Accessed 12 Sept., 2017. http://www.endofempire.asia/0809-2-the-day-after-4/.

CHAPTER 6

Living in Fear: Nuclear Films

Introduction

While the exhibition *The Family of Man* ultimately sought to visually reinforce the universality of the human experience, against the backdrop of the Cold War and the threat of nuclear war, the Japanese people also viewed a range of films that dealt more directly with nuclear issues in the 1950s. Anxieties about the atomic age jostled with dreams of how nuclear technology might one day offer a resource-poor Japan the tantalizing possibility of a safe and low-cost energy source that would underpin the nation's prosperity in the future.

It is clear that fears of nuclear technology informed Japan's science-fiction films during the 1950s. Films were produced which portray the US-Japan relationship and, indeed, the atom from differing perspectives. In this chapter, we will examine three documentaries and three fictional films produced in the space of a few years to show the differing views about nuclear technology, the significance of what happened to the crew of the Lucky Dragon in March 1954 and who was to blame, the state of the US-Japan relationship and what sort of future the world faces. All involved scientists. The films will include the American-made *The Beast from 20,000 Fathoms* (1953) which strongly influenced Godzilla films, *Shi no hai* (*Ashes of Death*) (1954) produced by Shin Riken Eigasha, *The Yukawa Story* (1954–1955) (USIS/USIA), *Gojira* (*Godzilla*) (1954) (Tōhō), *Ikimono no kiroku* (*I Live in Fear: Record of a Living Being*)

M. Low, *Visualizing Nuclear Power in Japan*, Palgrave Studies in the History of Science and Technology,
https://doi.org/10.1007/978-3-030-47198-9_6

(1955) (Tōhō) and *Ikite ite yokatta* (*Still, I'm Glad that I'm Alive*) (1956) (Gensuikyō and Japan Documentary Film Co.).

The Beast from 20,000 Fathoms (1953)

In June 1953, Eleanor Roosevelt visited Japan as part of a world tour. Her itinerary included a visit to Hiroshima. On 17 June, she met women at the Atomic Bomb Casualty Commission in Hiroshima who had been injured by the atomic bomb. In her newspaper column "May Day," she wrote of their wounds. Although sympathetic to the women, she echoed the official US government line when she wrote that many casualties were the result of fires that had started immediately after the dropping of the atomic bomb. There is no mention of radiation.[1] Just a few days earlier, the film *The Beast from 20,000 Fathoms* (1953) was released in the USA. There is similarly little direct reference in the film to radiation effects arising from an atomic bomb test conducted in the Baffin Bay area of the Arctic Circle. The bomb awakens a Rhedosaurus that has been sleeping deep below the ice for 100 million years. Two key characters in the film are the nuclear physicist Professor Thomas Nesbitt who works for the Atomic Energy Commission along with military liaison Col. John (Jack) Evans. Both are members of the scientific expedition Operation Experiment tasked with observing the test. Nesbitt and a colleague, Professor Ritchie, venture out to an observation post with Geiger counters in case of heavy radiation. But that is the last we hear of radiation until the very end of the film.

We eventually see the monster emerging from the water in lower Manhattan, against the backdrop of Brooklyn Bridge. It stomps through the city causing destruction. A state of emergency is declared, sirens sound and we are told that "This is full-scale war against a terrible enemy…such as modern man has never before faced." The monster that had been woken by the bomb now is seen as a symbol of it. The monster is located at Coney Island. Nesbitt believes that the only way to kill the monster without spreading its germs throughout a wide area is to shoot a radioactive isotope into the monster and destroy all its diseased tissue. In this way, viral contagion rather than radiation is depicted as the cause of loss of life. It also demonstrates how nuclear weapons can successfully be used to deal with the threat posed by other nuclear weapons (aka the Beast). It was no coincidence that radioactive isotopes were promoted in *Atoms for Peace* exhibits shown throughout the world as a peaceful spin-off of AEC nuclear programs that would contribute to medical science and human well-being.[2]

The Beast was released in Japan in 1954 just as the Japanese remake *Gojira* (*Godzilla*) (1954) (Tōhō) was screening in cinemas.[3] Despite the simultaneous showing of the two films in Japan, the Japanese film would be much more pronounced in its acknowledgement of the dangers of radiation.

Ashes of Death (1954)

The science documentary *Shi no hai* (*Ashes of Death*) (1954) which was also released in Japan that year also made clear to Japanese audiences the harmful effects of nuclear weapons. Produced by Shin Riken Eigasha and directed by Fueki Juzaburō, under the supervision of the scientist Kimura Kenjirō, *Ashes of Death* was just a short twenty-three-minute, three-reel, 16 mm film. It nevertheless attracted considerable attention, highlighting the plight of a tuna fishing boat, the Lucky Dragon in an incident that had occurred only in March of that year.

The company Shin Riken Eigasha was originally established in April 1938 as part of an industrial group associated with the Institute of Physical and Chemical Research (known in Japan as Riken for short) where Nishina Yoshio and other physicists such as Yukawa Hideki and Taketani Mituo had worked on what was part of Japan's small-scale atomic bomb project during World War II. Before the war, Nishina and Kimura had conducted research together and discovered a new radioactive isotope of uranium, ^{237}U, as well as symmetric nuclear fission.[4] By 1954, Kimura was Dean of the Faculty of Science, University of Tokyo, and well qualified to lead the radiochemical analysis of the white ashes of radioactive fallout that the crew of the Lucky Dragon had been exposed to.[5] He in turn provided scientific advice when the film *Ashes of Death* was made and presented his scientific findings at the 1955 International Conference on the Peaceful Uses of Atomic Energy, Geneva.[6]

A copy of the script of *Ashes of Death* was made available by Shin Riken Eigasha to the US government and found its way to George C. Spiegel of the Department of State's Office of the Special Assistant for Atomic Energy. The script provides a scene-by-scene description of the film, along with voice-over commentary.[7] The film begins by showing an idyllic scene of a peaceful sea. However, everything is rudely interrupted by the dropping of an atomic bomb. We witness a mushroom cloud, and a calendar indicates that it is 6 August 1945. Hiroshima is reduced to rubble and we see the wreckage of a tramcar. The film notes that Japan was the first nation to fall victim to the atomic bomb. We see the remains of dead men and women. The scene then moves to a mass commemoration ceremony for the atomic bomb victims in Hiroshima. The people urge "No More Hiroshimas!"[8]

We then see newspapers' headlines about the Lucky Dragon Incident and how twenty-three crew members of the tuna fishing boat had been severely burnt. The worst-affected cases were admitted to Tokyo University Hospital. We then see the physician Dr Tsuzuki Masao of Tokyo University write on a blackboard that the crew members were suffering from acute radiation sickness. We see more newspaper headlines about how tons of radioactive tuna had been buried in the ground and abandoned in the sea. Fears about possible contamination spread amongst housewives.[9] In this way, the documentary suggests a direct lineage from Hiroshima to the Lucky Dragon Incident and reinforces the belief that the Japanese people had fallen victim to the bomb a third time, this time a hydrogen bomb. We then see May Day demonstrations and voices calling for the prohibition of the hydrogen bomb.[10]

The film then turns to the Shunkotsumaru, the Fishery Board's training ship which left Shibaura, Tokyo, on 15 May 1954 destined for Bikini Atoll. On board were scientific experts who were seeking to collect radioactive materials in the hope that they might somehow assist in the treatment of the affected crew members of the Lucky Dragon No. 5. We next see Kimura Kenjirō announcing that twenty-six kinds of radioactive elements had been identified in an analysis of the radioactive fallout. Among the radioactive isotopes found was Strontium 90, radiation from which was known to be "the most dangerous and harmful for humans."[11]

The crew members are shown receiving medical treatment at Tokyo University Hospital. They are being allowed to rest, given blood transfusions and antibiotics but no remedy for their malady has been found. The film comments that this is against the backdrop of a race between the Americans and Russians to produce more powerful nuclear weapons. The House of Representatives passes a resolution calling for international control of atomic weapons. The film shows the physicist Yukawa Hideki calling for a stop to nuclear weapons research and we see Tsuzuki Masao of Tokyo University making clear that the world's scientists had yet to arrive at a remedy for radiation sickness.[12]

We hear other voices as well. At a meeting of the South-Eastern Conference of the World Pacifists' Association held in Japan, participants call for a total ban on all nuclear weapons. We see people praying and the film reminds us that the Japanese people have experienced the tragic consequences of nuclear blasts. We see traces of burns on young Japanese girls. The film hopes that there are no more such tragedies awaiting the Japanese people. It is in the final moments of the film that we are told that atomic power should only be used for peaceful purposes that contribute "brightly" to the prosperity of nations and the welfare of human beings throughout the world. A girl is shown praying and the words of Yukawa Hideki effectively calling for more social responsibility among scientists appear on the screen. The endless frontier of science can sometimes lead to unexpected results that could in turn lead to mass destruction.[13]

Gojira (*Godzilla*) (1954)

In the film *Gojira* (*Godzilla*) (1954), science is portrayed as initially bad. Nuclear tests had awakened the prehistoric creature Godzilla from his slumber at the bottom of the ocean and exposed him to radiation, giving him his monstrous form. Godzilla is, like the Japanese people themselves, the victim of radiation exposure. His thick and furrowed skin is not unlike the scarred skin of atomic bomb victims. At the same time, his capacity to cause havoc in Tokyo shows the destructive capabilities of things nuclear.[14]

The film had clearly been influenced by *The Beast from 20,000 Fathoms* and its initial working title is said to have been *The Giant Monster from 20,000 Miles beneath the Sea*.[15] In the Japanese film, Tokyo takes the place of New York City. A prehistoric creature stomps through a major city causing destruction in both films. In *The Beast*, a palaeontologist Professor Elson advocates capturing it alive and studying it. Likewise in *Godzilla*, the Japanese palaeontologist Yamane Kyōhei does not want to kill the monster.

Yamane's daughter, Emiko, faces the moral dilemma of whether or not to reveal the secret weapons that her fiancé, Serizawa Daisuke, has been working on. Should she maintain scientific confidentiality or reveal a weapon that could potentially save many lives? She ultimately betrays Serizawa by telling her real love Ogata Hideto from Southern Sea Salvage Company, what she had observed in Serizawa's laboratory.

Unlike the Beast, Godzilla is a mutant monster that emits radiation.[16] The threat posed by radiation is much more strongly emphasized in the

Japanese film. But seen more broadly, we can say that in both films, science does triumph. In *The Beast* and *Godzilla*, science would ultimately come to the rescue but in the latter, it would be a somewhat pyrrhic victory. Like Nesbitt with his radioactive isotope, the scientist Serizawa Daisuke seemingly destroys Godzilla with an oxygen destroyer that he had secretly created. Yet whereas the message from the American film seems to be that nuclear science can solve the problems that it creates, the Japanese take on it is that nuclear weapons are wrong and that there is a moral responsibility not to encourage the proliferation of weapons of mass destruction, whether they be atomic and hydrogen bombs or the "oxygen destroyer". Serizawa kills himself and takes the secret of the oxygen destroyer with him.[17] The public responded strongly to the film with over 10 per cent of the Japanese population estimated as having seen the film.[18]

In contrast to the Beast which is depicted as a malevolent monster which preys on people, Godzilla is portrayed as a somewhat more sympathetic creature whose past has been a part of Ōdo Island folklore. The locals have long feared Godzilla and have sought to placate it through sacrifices and ceremonies. Unfortunately for them, hydrogen bomb testing has altered Godzilla's natural habitat and the creature has fallen victim to a massive dose of radiation which renders it radioactive. Despite that, Godzilla has managed to survive and, in the film, scientists like Yamane are curious about how it has survived a hydrogen bomb rather than wanting to know how best to kill it. Godzilla destroys parts of Tokyo, but it has been pointed out that the monster does not seem to specifically chase after and target human beings.[19] It is differences like these that show how the Japanese, in this case director Honda Ishirō, has adapted *The Beast* to depict the Japanese response to hydrogen bomb testing and more broadly nuclear fear.

Yamane reports to parliament that Geiger counter readings show that Godzilla must have absorbed a massive dose of radiation from a hydrogen bomb test. People quickly come to the conclusion that the monster is a product of nuclear weapons. There is a debate about whether or not to make the information about Godzilla public in case the whole country panics but there is a strong call to let the people know the truth. Newspapers report that Counter-Godzilla Headquarters have been established but already seventeen ships had been lost in efforts to stave off Godzilla.

It is against this background that the film shows a scene of commuters on a train in Tokyo who discuss the recent terrible events. Not only had there been a nationwide scare about contaminated tuna and concern for the crew of the Lucky Dragon No. 5 tuna fishing boat being affected by radioactive fallout but also now there was Godzilla to contend with. They joke that Godzilla will eat the woman in the group in one bite. They lament how they will have to seek out a shelter to protect themselves in the case of Godzilla rising up from Tokyo Bay. They think back to their experiences during the war when they had to do the same and seek refuge from attack.

While *The Beast from 20,000 Fathoms* showed Japanese film makers how to express nuclear anxieties through a monster movie, it is somewhat ironical that Godzilla would be reimported back to the USA and released in Americanized form as *Godzilla, King of the Monsters!* (1955) and spawn many sequels in both Japanese and English. The original film's success contributed to the rise of the *tokusatsu* genre of live actions films that make extensive use of special effects and often include superheroes. Over time the genre has tended to downplay the dangers of radiation exposure and allowed the Japanese to "ease into the commercial use of nuclear power as an energy sources."[20] The nuclear fears of the Japanese people are projected on to the bodies of monsters and aliens who fight it out. By doing so, the audience can detach themselves from their concerns about things nuclear. Ten years after the release of the first Godzilla film, Honda Ishirō's *Ghidorah, the Three Headed Monster* (1964) would transform Godzilla into a hero who would work with Mothra and Rodan to save the world from Ghidorah. It showed that, through careful management, Japan could use nuclear power for beneficial purposes.

The Yukawa Story (1954–1955)

The Yukawa Story: As Told by Taka-aki Yukawa (1954–1955) was produced in the USA by the USIS/USIA produced in the hope of promoting better bilateral relations and portraying science in a more positive light. It was released in Japan as *Chichi Yukawa Hakase* (*My Father, Dr Yukawa*).[21] The story is told from the perspective of Yukawa Taka-aki, one of two sons of the Nobel Prize-winning physicist Yukawa Hideki. The second son Harumi, who was one year older than Taka-aki, does not appear in the film. The film is told through Taka-aki's eyes. We see him walking across Brooklyn Bridge and gazing across at the city. He refers to his father as Dr Yukawa.

We get a sense of the world of science that Yukawa Hideki is immersed in. We see him at work at the Cosmotron at Brookhaven National

Laboratory near New York City. The Cosmotron was a particle accelerator, a proton synchrotron that had just reached its full energy in 1953.[22] It was used to observe mesons previously only seen in cosmic rays.[23] Yukawa had predicted the existence of the meson. We then see Yukawa in his office at Columbia University, New York City. This is contrasted with the Yukawa home in the ancient capital of Kyoto where he was director of the newly established Yukawa Hall at Kyoto University. The university helped organize the International Conference of Theoretical Physics that was held in Kyoto and Tokyo in September 1953. We also see film footage of Yukawa at the Institute for Advanced Study at Princeton University c. 1948 where he had been invited by J. Robert Oppenheimer, director of the Institute to visit for a year. Yukawa was subsequently appointed professor at Columbia and awarded the Nobel Prize in Physics in 1949 for his work on the hitherto unknown subatomic particle, the meson which is considered responsible for the forces that hold the nucleus together. The film returns to Yukawa in his office at Pupin Hall at Columbia, and we see the cover of a scientific paper on meson theory.

We next see Dr Yukawa with his wife Sumi and son Taka-aki. There is a flashback to Mrs Yukawa performing a traditional Japanese dance, Musume Dōjōji (The Maiden at Dōjōji Temple) on a stage with some simple props: a hanging bell and a tree in bloom. The bell is lowered. There is a costume change. The scene highlights Mrs Yukawa's interest in the arts. In this way, Taka-aki's parents are characterized as representing both the past and future, and their son's dilemma is which to choose—return to his homeland and the past or to pursue the study of science in the USA.

The film returns to the Brooklyn Bridge. Taka-aki is deliberating over whether he should follow the example of his father (a theoretical physicist) or that of his mother (a practicing Buddhist). He graduates from the Bronx High School of Science and he needs to make a decision: how to reconcile his father's world, the world of science and inductive reasoning, with his mother's more traditional one which centred on an "emotional" appreciation of cultural heritage. But the film suggests that he can choose a union of both worlds, like the marriage of his parents, rather than seeking to separate the two.

Yukawa is shown inspecting a 10 MeV cyclotron in the basement at Pupin Hall at Columbia University with an American professor and student. The cyclotron had been one of the first particle accelerators built in the USA. It was used in 1939 to confirm the splitting of uranium atoms that had just been reported in Europe. In the following year, it was crucial in helping scientists to determine that ^{235}U (uranium-235) could more readily split into two. This led the way to studies into the possibility of chain reactions.[24] Cyclotrons were later also used to artificially produce mesons. The film explains that

> when you leave the great magnet of the cyclotron and take back once again your watch, you come from a place where time is measured by the ten thousandth part of a second to our own everyday kind of time. With the human problems of this our own time, a machine cannot help us.

The film transitions to a scene showing Yukawa Hideki and his wife visiting a local Buddhist temple. We are told that "to settle the difficulties that man has with himself, we must look to a different kind of wisdom." Taking a conciliatory tone, the film rationalizes things in the following way:

> We need the freedom of the West to progress and expand. We need the Oriental acceptance of tradition to know the responsibility we all share as human beings. If out of these two civilizations, a new synthesis can be achieved, there is a chance for growth. If not…[an ominous sound is heard]

In this way, the threat of nuclear war is intimated.

Soon it is time for the Yukawas to leave New York City. Taka-aki comes to realize that both worlds come together in his parents' lives and that it is a not a choice of one or the other. Rather, it is something more difficult. He decides to continue his education in the USA where he is able to "combine the knowledge he acquires from the U.S. and his Japanese cultural heritage to better serve his own country."[25] He resolves to stay behind in the USA while his parents return to Japan. Taka-aki sees his parents off at the airport, holding back tears. In this way, the film shows how the USA complemented Japan, a relationship not unlike that of the Yukawas themselves. It was seen as helping to counter anti-American feeling engendered by the Lucky Dragon Incident and a way of promoting US-Japan ties.

This is despite Yukawa having been personally shocked by the Lucky Dragon Incident and concerned about the danger of nuclear weapons.[26] In the film, however, the spectre of nuclear war is very much in the background and hardly alluded to. Nor is there any reference to Yukawa's involvement in Japan's own small-scale wartime atomic bomb project. This included the "F Project" (F for "fission") led by the physicist Arakatsu

Bunsaku at Kyoto Imperial University and sponsored by the Imperial Japanese Navy in the last stages of the war (c. mid-1943–1945). Yukawa participated in project meetings in 1945 and was apparently responsible for nuclear theory.[27]

Yukawa's 1949 Nobel Prize in physics in 1949 and his status as the first Japanese Nobel laureate helped to establish Japanese claims to scientific prowess. Given that State Shintō and Japan's warrior past had been discredited after the war, science was seen as helping to build a new Japan. Scientific ability would increasingly become a marker of national character. Even in Japanese films that addressed nuclear fear such as *Godzilla*, there is hope that Japanese ingenuity in science might save the Japanese people.

I Live in Fear (1955)

In contrast to the relative optimism of *The Yukawa Story*, Kurosawa Akira would direct the 1955 film *Ikimono no kiroku* (*I Live in Fear: Record of a Living Being*), arguably one of the most well-known examples of Japanese fear of a nuclear attack. The film begins with the opening credits superimposed on a busy street scene with a high-pitched screechy soundtrack composed by Hayasaka Fumio who was dying from tuberculosis at the time. There are jazzy inflections from a saxophone, the clashing of cymbals and a zither-like sound reminiscent of a 1950s science-fiction film.[28] One cannot help but imagine that a flyer saucer is hovering over the city! This complex cacophony of sounds conveys the emotional intensity of the nuclear fears of the principal character, Nakajima Kiichi. The noise gives way to the ordinary sounds of the street and leads to the opening scene in the dental clinic of Dr Harada who also works as a court mediator. He is asked to assist in resolving a family dispute centring on Nakajima and his fears. Although Harada is not a scientist per se, the film portrays him as keen to be objective and assess the dispute from both sides.

The director Kurosawa avoids being too political and portrays the bomb in terms of its psychological force on individuals and their families. The owner of a foundry, Nakajima, seeks to escape the "ashes of death" of nuclear tests being conducted in the Pacific and the threat of possible nuclear war by migrating to Brazil where he also hopes to take his extended family. The family resist and in an attempt to protect their inheritance, resort to going to the Tokyo Family Court to have Nakajima declared mentally incompetent.[29]

Unlike the film *Godzilla*, there is no visible monster or alien, only fears surrounding the threat to human life posed by nuclear devastation. The monster is thus within. It cripples Nakajima, an aging industrialist. Like Nakajima, the Japanese public were reading about nuclear tests and radioactive fallout, and were increasingly aware of the debates about the introduction of nuclear power and how best to go about it. But unlike him, they went about their everyday life. Nakajima is so paralysed with anxiety that the sound of a plane, thunder and lightning cause him to cower in fear as they seem to him to mimic a nuclear attack.[30] Nakajima ultimately sets light to the family foundry in the hope that it will force the family to migrate with him. He eventually goes insane, thinking that he is on another planet and that the earth is on fire.[31]

One cannot help but ask whether Nakajima is insane or whether the Japanese people themselves are the deluded ones lulling themselves into a false sense of security. Throughout the film we see characters wiping away sweat and fanning themselves in an attempt to deal with the unbearable heat. This seems to be a metaphor for how the Japanese people have normalized nuclear fear and go about their lives attempting to deal with the stress associated with everyday life, powerless to act on their fears.

There is little overt political commentary about the nuclear weapon race and the Cold War. We do see Harada reading a book entitled *Shi no hai* (*Ashes of Death*) and Nakajima tears a newspaper depicting a hydrogen bomb test, but we see no reference to the Japanese movement to ban nuclear weapons. At the end of his review of the film in late 1955, the physicist Taketani Mituo wrote that more reference to the present status and nature of the movement would have strengthened the film.[32] Rather than being overwhelmed by their fears of nuclear devastation, he and others advocated acting on their fears and lending their voices to the movement calling for the prohibition of atomic and hydrogen bombs.[33]

Nakajima had only become very fearful of nuclear weapons from June 1954, in the wake of the Lucky Dragon Incident. He, like many other Japanese, had developed a belated fear of things nuclear. Nakajima is not afraid of dying but does not wish to be killed. He is determined to avoid nuclear weapons if he can help it. In this regard, he seems to reflect the thinking of many people who joined the movement to ban the bomb.[34] Of course, while many Japanese might fear the bomb, most would not have the resources to escape it nor the desire to leave their home and country.[35] In a way, the film also addresses the tendency of the Japanese public to accept nuclear weapons as inevitable and to suppress any fears that they might have regarding them.

Still, I'm Glad that I'm Alive (1956)

The year after the release of Kurosawa's film, Kamei Fumio's documentary *Ikite ite yokatta* (*Still, I'm Glad that I'm Alive*) (1956) was released. The forty-eight-minute, 16 mm film begins by showing the damaged face of a statue at Urakami Cathedral at Nagasaki. We then see images of the new peace bridge at Hiroshima, the A-Bomb Dome and a stray dog among the ruins. The film is notable not only for showing footage of the destruction at Hiroshima and Nagasaki soon after the atomic bomb had been dropped, but also for turning the camera to the *hibakusha* (atomic bomb survivors) themselves, their injuries and their daily struggle to survive ten years later.

The film was made against the backdrop of the Lucky Dragon Incident and the movement to ban nuclear weapons. It was produced by Gensuikyō (Japan Council against Atomic and Hydrogen Bombs) and the Japan Documentary Film Co. The roots of Gensuikyō can be traced to a group of housewives in Suginami ward in Tokyo. The women participated in a book club led by a former Tokyo University professor Yasui Kaoru. The club circulated a local petition to ban the bomb and in fifty days gathered an impressive 270,000 signatures. This gave rise to a nationwide movement and millions of Japanese signed the petition.[36] The film which depicts a number of stories of *hibakusha* and their lives was produced in order to commemorate the Council's first international conference held in Hiroshima in 1955.[37] Halfway through the film, we see an atomic bomb survivor speak at the conference about her experience.

Much earlier in the film, we are shown a map showing the location of Enewetak and Bikini Atolls in the Marshall Islands, and then the fiery blast of the hydrogen bomb. We see crew members of the tuna fishing boat Lucky Dragon No. 5 receiving medical treatment at Tokyo University Hospital. We see a jar of radioactive fallout and hear the ominous words "*Shi no hai*" ("*Ashes of Death*"). Crew member, Kuboyama Aikichi, is shown lying in bed and then we see his wife and her mother who are both crying. This is followed by footage of his draped coffin. He had become the first Japanese victim of the hydrogen bomb test.

Conclusion

These films, all produced within a few years of each other, show how Japanese awareness of nuclear issues was heightened immediately after the Lucky Dragon Incident. Like US-sponsored exhibitions such as *Atoms for Peace* and *The Family of Man* that were shown in Japan, *The Beast from 20,000 Fathoms* would also provide an American template for how to depict nuclear fear. But the Lucky Dragon Incident would shape Godzilla and ensure that Japanese monster films would not be mere remakes. Popular representations of the threat of nuclear weapons remind us of how hard proponents of nuclear power had to work to ensure that the Japanese people could embrace atoms for peace and at the same time call for bans on nuclear weapons. In Japan, like elsewhere, the atomic age seemed to be like a double-edged sword. For the Japanese who had experienced the atomic bomb at Hiroshima and Nagasaki, there seemed to be an option that they could say no to the atomic bomb and only take the good, namely civilian nuclear power. But this hid the reality that post-war Japan enjoyed the benefit of the US nuclear umbrella.

Notes

1. Eleanor Roosevelt, "My Day, June 17, 1953," *The Eleanor Roosevelt Papers Digital Edition* (2017), accessed 17 May 2019, https://www2.gwu.edu/~erpapers/myday/displaydoc.cfm?_y=1953&_f=md002565
2. Alexander Hammond, "Rescripting the Nuclear Threat in 1953: *The Beast from 20,000 Fathoms*," *Northwest Review* 22, no. 1 (1 Jan. 1984): 181–194, esp. 190.
3. Takayuki Tatsumi, trans. Seth Jacobowitz, "On the Monstrous Plant, Or How Godzilla Took a Roman Holiday," in *The Liverpool Companion to World Science Fiction Film*, ed. Sonja Fritzche (Liverpool: Liverpool University Press, 2014), 69–85, esp. 77.
4. Nagao Ikeda, "The Discoveries of Uranium 237 and Symmetric Fission: From the Archival Papers of Nishina and Kimura," *Proceedings of the Japan Academy, Series B, Physical and Biological Sciences* 87, no. 7 (2011): 371–375.
5. H. K. Yoshihara, "The Dawn of Radiochemistry in Japan," *Radiochimica Acta* 100 (2012): 523–527.
6. Yoshihara, "The Dawn of Radiochemistry in Japan."
7. "Ash of Death" script, U.S. Department of State, General Records, RG 59, Office of the Secretary, Special Assistant to Secretary of State for

Atomic Energy and Outer Space, General Records Relating to Atomic Energy Matters, 1948–1962, Country File: Japan, U.S. National Archives and Records Administration, College Park, MD.

8. "Ash of Death" script, 2.
9. "Ash of Death" script, 3.
10. "Ash of Death" script, 4.
11. "Ash of Death" script, 4–5.
12. "Ash of Death" script, 8.
13. "Ash of Death" script, 9.
14. Yuki Miyamoto, "Gendered Bodies in *Tokusatsu*: Monsters and Aliens as the Atomic Bomb Victims," *Journal of Popular Culture* 49, no. 5 (2016): 1086–1106, esp. 1091.
15. William Tsutsui, *Godzilla on My Mind: Fifty Years of the King of Monsters* (New York: Palgrave Macmillan, 2004), 20.
16. Yuki Tanaka, "Godzilla and the Bravo Shot: Who Created and Killed the Monster?" *Asia-Pacific Journal* 3, no. 6 (10 Jun. 2005), https://apjjf.org/-Yuki-Tanaka/1652/article.pdf
17. John Vohlidka, "Atomic Reaction: Godzilla as Metaphor for Generational Attitudes toward the United States and the Bomb," in *The Atomic Bomb in Japanese Cinema: Critical Essays*, ed. Matthew Edwards (Jefferson, North Carolina: McFarland and Co., 2015), 56–67, esp. 58.
18. Miyamoto, "Gendered Bodies in *Tokusatsu*."
19. Tanaka, "Godzilla and the Bravo Shot."
20. Miyamoto, "Gendered Bodies in *Tokusatsu*," 1088.
21. Tsuchiya, Yuka, "The Atoms for Peace USIS Films: Spreading the Gospel of the "Blessing" of Atomic Energy in the Early Cold War Era," *International Journal of Korean History* 19, no. 2 (Aug. 2014): 107–135, esp. 124–125.
22. L.D. Stoughton, "The Cosmotron Building," *Review of Scientific Instruments* 24, no. 9 (Sept. 1953): 854–855; M. Hildred Blewett, "The Cosmotron: A Review," *Review of Scientific Instruments* 24, no. 9 (Sept. 1953): 725–737.
23. W.B. Fowler, R.P. Shutt, A.M. Thorndike and W.L. Whittemore, "Examples of Multiple Pion Production in *n-p* Collisions Observed at the Cosmotron," *Physical Review* 91, no. 3 (Aug. 1953): 758–759.
24. William J. Broad, "Columbia's Historic Atom Smasher Is Now Destined for the Junk Heap," *New York Times*, 20 Dec. 2007, B1.
25. "Report Prepared by the National Security Council," 2 Mar. 1955, Washington D.C. In U.S. Department of State, *Foreign Relations of the United States, 1955–1957, Foreign Policy; Foreign Information Program, Vol. IX*, ed. John P. Glennon (Washington, D.C.: U.S. Government Printing Office, 1987), 517.

26. Akira Kurosaki, "Japanese Scientists' Critique of Nuclear Deterrence Theory and Its Influence on Pugwash, 1954–1964," *Journal of Cold War Studies* 20, no. 1 (Winter 2018): 101–139, esp. 111.
27. "Logs of Nobel Prize-winning Physicist Yukawa Shows Clues on Wartime Nuke Research," *Mainichi Newspapers*, 22 Jul. 2018, https://mainichi.jp/english/articles/20171222/p2a/00m/0na/028000c, accessed 25 May 2019; Morris Low, *Science and the Building of a New Japan* (New York: Palgrave Macmillan, 2005), 38.
28. James Goodwin, *Akira Kurosawa and Intertextual Cinema* (Baltimore: Johns Hopkins University Press, 1994), 231.
29. Tadao Satō, *Currents in Japanese Cinema*, trans. by Gregory Barrett (Tokyo: Kodansha International, 1982, 1987), 129, 199–200.
30. James Goodwin, "Akira Kurosawa and the Atomic Age," in *Perspectives on Akira Kurosawa*, ed. James Goodwin (New York: G.K. Hall and Co., 1994), 124–142, esp. 131.
31. Satō, *Currents in Japanese Cinema*, 129, 199–200.
32. Taketani, Mituo, "Shōchōshugi no genkai" ("The Limits of Symbolism"), *Kinema junpō* (*Motion Picture Times*), no. 133 (Dec. 1955): 49–50, esp. 50.
33. Yamamoto, Akihiro, *Kaku to Nihonjin* (*Japanese and the Nuclear*) (Tokyo: Chūō Kōronsha, 2015), 33–34.
34. Taketani, "Shōchōshugi no genkai," 49.
35. Satō, *Currents in Japanese Cinema*, 129, 199–200.
36. George O. Totten and Tamio Kawakami, "Gensuikyō and the Peace Movement in Japan," *Asian Survey* 4, no. 5 (May 1964): 833–841, esp. 834–835.
37. Yuko Shibata, "Belated Arrival in Political Transition: 1950s Films on Hiroshima and Nagasaki," in *When the Tsunami Came to Shore: Culture and Disaster in Japan*, Roy Starrs (Leiden: Global Oriental, 2014), 231–248, esp. 235–238. For commentary on the conference, see Lawrence S. Wittner, *The Struggle against the Bomb*, Vol. 2 (Stanford, Calif.: Stanford University Press, 1997), 9–10.

Bibliography

Blewett, M. Hildred. "The Cosmotron: A Review." *Review of Scientific Instruments* 24, no. 9 (Sept. 1953): 725–737.

Broad, William J. "Columbia's Historic Atom Smasher Is Now Destined for the Junk Heap." *New York Times*, 20 Dec. 2007.

Fowler, W.B., R.P. Shutt, A.M. Thorndike and W.L. Whittemore. "Examples of Multiple Pion Production in *n-p* Collisions Observed at the Cosmotron." *Physical Review* 91, no. 3 (Aug. 1953): 758–759.

Goodwin, James. *Akira Kurosawa and Intertextual Cinema*. Baltimore: Johns Hopkins University Press, 1994a.

Goodwin, James. "Akira Kurosawa and the Atomic Age." In *Perspectives on Akira Kurosawa*. Edited by James Goodwin, 124–142. New York: G.K. Hall and Co., 1994b.

Hammond, Alexander. "Rescripting the Nuclear Threat in 1953: *The Beast from 20,000 Fathoms*." *Northwest Review* 22, no. 1 (1 Jan. 1984): 181–194.

Ikeda, Nagao. "The Discoveries of Uranium 237 and Symmetric Fission: From the Archival Papers of Nishina and Kimura" *Proceedings of the Japan Academy, Series B, Physical and Biological Sciences* 87, no. 7 (2011): 371–375.

Kurosaki, Akira. "Japanese Scientists' Critique of Nuclear Deterrence Theory and Its Influence on Pugwash, 1954–1964." *Journal of Cold War Studies* 20, no. 1 (Winter 2018): 101–139.

"Logs of Nobel Prize-winning Physicist Yukawa Shows Clues on Wartime Nuke Research." *Mainichi Newspapers*, 22 Jul. 2018. Accessed 25 May, 2019 https://mainichi.jp/english/articles/20171222/p2a/00m/0na/028000c.

Low, Morris. *Science and the Building of a New Japan*. New York: Palgrave Macmillan, 2005.

Miyamoto, Yuki. "Gendered Bodies in *Tokusatsu*: Monsters and Aliens as the Atomic Bomb Victims," *Journal of Popular Culture* 49, no. 5 (2016): 1086–1106.

"Report Prepared by the National Security Council," 2 Mar. 1955, Washington D.C. In U.S. Department of State, *Foreign Relations of the United States, 1955–1957, Foreign Policy; Foreign Information Program, Vol. IX*. Edited by John P. Glennon. Washington, D.C.: U.S. Government Printing Office, 1987.

Roosevelt, Eleanor. "My Day, June 17, 1953." *The Eleanor Roosevelt Papers Digital Edition* (2017). Accessed 17 May, 2019. https://www2.gwu.edu/~erpapers/myday/displaydoc.cfm?_y=1953&_f=md002565.

Satō, Tadao. *Currents in Japanese Cinema*. Translated by Gregory Barrett. Tokyo: Kodansha International, 1982, 1987.

Shibata, Yuko. "Belated Arrival in Political Transition: 1950s Films on Hiroshima and Nagasaki." In *When the Tsunami Came to Shore: Culture and Disaster in Japan*. Edited by Roy Starrs, 231–248. Leiden: Global Oriental, 2014.

Stoughton, L.D. "The Cosmotron Building." *Review of Scientific Instruments* 24, no. 9 (Sept. 1953): 854–855

Taketani, Mituo, "Shōchōshugi no genkai" ("The Limits of Symbolism"). *Kinema junpō* (*Motion Picture Times*), no. 133 (Dec. 1955): 49–50.

Tanaka, Yuki. "Godzilla and the Bravo Shot: Who Created and Killed the Monster?" *Asia-Pacific Journal* 3, no. 6 (10 June 2005), https://apjjf.org/-Yuki-Tanaka/1652/article.pdf.

Tatsumi, Takayuki, translated by Seth Jacobowitz. "On the Monstrous Plant, Or How Godzilla Took a Roman Holiday." In *The Liverpool Companion to World*

Science Fiction Film. Edited by Sonja Fritzche, 69–85. Liverpool: Liverpool University Press, 2014.

Totten, George O. and Tamio Kawakami. "Gensuikyō and the Peace Movement in Japan." *Asian Survey* 4, no. 5 (May 1964): 833–841.

Tsuchiya, Yuka. "The Atoms for Peace USIS Films: Spreading the Gospel of the "Blessing" of Atomic Energy in the Early Cold War Era." *International Journal of Korean History* 19, no. 2 (Aug. 2014): 107–135.

Tsutsui, William. *Godzilla on My Mind: Fifty Years of the King of Monsters*. New York: Palgrave Macmillan, 2004.

US National Archives and Records Administration, College Park, MD. RG 59. U.S. Department of State, General Records.

Vohlidka, John. "Atomic Reaction: Godzilla as Metaphor for Generational Attitudes toward the United States and the Bomb." In *The Atomic Bomb in Japanese Cinema: Critical Essays*. Edited by Matthew Edwards 56–67. Jefferson, North Carolina: McFarland and Co., 2015.

Wittner, Lawrence S. *The Struggle against the Bomb*, Vol. 2. Stanford, Calif.: Stanford University Press, 1997.

Yamamoto, Akihiro. *Kaku to Nihonjin* (*Japanese and the Nuclear*). Tokyo: Chūō Kōronsha, 2015.

Yoshihara, H. K. "The Dawn of Radiochemistry in Japan." *Radiochimica Acta* 100 (2012): 523–527.

CHAPTER 7

Making Atomic Dreams Real: 1956–1958

Introduction

In this chapter, we examine the dreams entertained by Japanese regarding atoms for peace and their efforts to realize them quickly through the purchase of reactors from abroad. Given the close contact between the USA and Japan since the American-led Allied Occupation, there was the expectation that Japan would turn to the USA when purchasing their first nuclear power plant. The visit by Sir Christopher Hinton from Great Britain's Atomic Energy Authority in 1956 was pivotal in persuading Shōriki Matsutarō, Chairman of Japan's Atomic Energy Commission to opt for an improved version of the British Calder Hall type power plant. Thanks to USIS effort to promote the public understanding of the peaceful use of atomic energy through touring exhibits, a narrative emerged that incorporated atomic energy as part of Japan's reconstruction and Japan's future. Expo '58, known as the "Atomic Fair," provided an opportunity for Japan to project this narrative to the rest of the world, against the backdrop of great international interest in the possibilities opened up by the development of atomic energy.

M. Low, *Visualizing Nuclear Power in Japan*, Palgrave Studies in the History of Science and Technology,
https://doi.org/10.1007/978-3-030-47198-9_7

Dream of Atoms for Peace

On 1 January 1956, *Nippon Times* staff writer Kurihara Koichirō predicted that Shōriki Matsutarō, owner of the *Yomiuri Shimbun* newspaper, Minister of State and Chairman of the Japan Atomic Energy Commission, would likely be Man of the Year in terms of the person who would have most influence on the destiny of Japan in the coming year. It was from that day that the JAEC would formally come into being.[1]

Shōriki also was part of a round-table discussion on "The Dream of Atoms for Peace" which was reported on in the pages of the *Yomiuri* that same day. Other participants included Liberal Democratic Party politician Nakasone Yasuhiro who was chairman of the government's Joint Committee for Atomic Energy, the physicist Sagane Ryōkichi, the female novelist Morita Tama, and *Yomiuri* editor Kojima.[2] There was a perception that in 1956, there would be real progress in dealing with atomic energy issues.

Kojima began by remarking on how the *Atoms for Peace* exhibition at Hibiya Park in 1955 had been particularly notable for giving the younger generation the hope and dream that atomic energy would help bring happiness to humankind, making the impossible possible. Sagane stated that seven or eight years ago, he had the fanciful hope that atomic energy might one day enable humankind to control the climate! Fortunately for him, at the 1955 Geneva Conference, the physicist Edward Teller, known as the father of the hydrogen bomb, had said that the bomb could be used in a controlled way to change the climate so his dream was not so ridiculous after all![3]

Nakasone's first dream was to use atomic energy to create radioactive spas throughout Japan. Ever the populist, he hoped that Japan would quickly construct reactors and produce cheap radioisotopes that could then be distributed to public bath houses throughout Japan so that the masses could enjoy bathing in radioactive spas that were ten times stronger than existing radium spas, safe and low in cost. In addition to the dream of using atomic energy to generate electric power, he also entertained hopes that such power could be used to operate desalination plants that could transform the abundant supplies of sea water into both freshwater and salt which Japan continued to have to import.[4]

Morita suggested that one problem was that some people could not separate atomic energy from the bomb and wanted to keep their distance. She thought that it was necessary to clearly distinguish between the two.

It was not clear what peaceful atomic industries were. It was necessary to make people aware of how atomic energy could make people's lives better.[5] Young women in particular were concerned about radioactivity. There was the perception that even if atomic energy was used for peaceful industrial purposes, as long as the threat posed by dangerous levels of radiation remained, the attitude was that it would not result in progress for humankind and opposition to nuclear power would remain.[6] Morita's contribution to the round-table discussion shows the highly gendered nature of responses to nuclear power in Japan.

It was men who often played a leading role in promoting civilian nuclear power and it was women who tended to air their concerns. This reflected the division of labour in postwar Japanese families. Men would often be employed as "salary men" in white collar office jobs while women took on the role of housewife and providing care for the family. As Kano Mikiyo puts it, men became associated with "progress and development"[7] and women were left to deal with its social consequences. Indeed, it was the protests of housewives in Suginami ward in Tokyo in response to the Lucky Dragon Incident in 1954 and their calls for an end to nuclear weapons testing that eventually became a nationwide movement against atomic and hydrogen bombs.[8] This led to the first World Conference against Atomic and Hydrogen Bombs in Hiroshima in 1955 and the second in Nagasaki in 1956. Despite all this, the *Atoms for Peace* exhibition continued to tour in 1956 and some of the exhibits would be donated to the Hiroshima Peace Memorial Museum the year after, when the tour had come to an end.[9]

Atoms for Peace Exhibition in Nagoya, Osaka and Hiroshima

The year 1956 was a big year for promoting the peaceful uses of atomic energy. The exhibition opened at the newly-built Aichi Prefectural Museum of Art in Nagoya on 1 January after having travelled from Tokyo. The Art Museum itself had been built to commemorate the signing of the San Francisco Peace Treaty as part of the Aichi Prefecture Cultural Centre and had opened in February 1955.[10] So it was an appropriate venue in terms of promoting US-Japan relations and the modernity that the Atoms for Peace campaign was seeking to associate itself with.

After Nagoya, the exhibition travelled to Kyoto and Osaka. At Osaka, it was held from late March through to April 1956 for some six weeks at the Asahi Ice Arena which was located at the edge of the downtown area. It complemented the Third Japan International Trade Fair which was also being held in Osaka, 8–22 April 1956. The fair was spread over two venues and the US pavilion was located on the main venue in a park area near the waterfront. There were twenty-six separate displays by prominent American companies, all housed within a pillar-less modernistic building with a canvas top that was likened to a large slice of red and white cake.[11] The structure included a 250-seat theatre where documentary films were shown and General Electric Co. put on their "House of Magic" live exhibition.[12] The latter had been a GE feature since the 1939 New York World's Fair. Science was presented as a series of magic tricks which engaged audiences in the USA and elsewhere.[13]

American companies were spruiking for business in Japan with Japan being second only to Canada as a buyer of American products.[14] Within the US pavilion, there was a special lounge where members of a US Department of Commerce trade mission to Japan were available to meet with Japanese business men to discuss US policy and trade practices.[15] This enthusiasm for trade with the Japanese was also partly fuelled by Cold War tensions. A CIA intelligence report produced the following month noted how the Sino-Soviet Bloc was participating in international trade fairs as a vehicle for propaganda and a way of implementing the Bloc's foreign economic policy.[16] East Germany participated in the Osaka fair where it was represented by a Carl Zeiss booth showing optical equipment.[17]

After Osaka, the *Atoms for Peace* exhibition arrived in Hiroshima where it clocked up its one-millionth visitor in Japan. It reached that milestone on 1 June, less than a week after it had opened in Hiroshima on 27 May. A fifteen-year-old, male high school student named Sakai Akira was the recipient of all the attention. A group of officials from the City of Hiroshima, Hiroshima prefecture, Hiroshima University, *Chūgoku Shimbun* newspaper and the American Cultural Centre were all there to greet him, in their capacity as sponsors of the exhibition. The American Cultural Centre was represented by its director, Abol Fazl Fotouhi. Young Sakai was shown through the exhibition by the officials who presented him with a large, red and white shoulder band which served to identify him as the one-millionth visitor. His school received a television set courtesy of a local store, reinforcing the perception that atomic energy would help the Japanese lead better lives. During the tour of the exhibition, young Sakai was presented a bouquet of flowers by the remotely controlled "magic hands". They even wrote out his name in Chinese characters and Sakai himself had an opportunity to operate the mechanical arms and hands.[18]

The happy occasion hid from view all the drama behind the holding of the exhibition in the Hiroshima Peace Memorial Museum. As early as December 1954, USIS officials had considered holding the exhibition first in Hiroshima in 1955, rather than Tokyo. But the proposal to do so was rejected as it would have made the link between civilian nuclear power and nuclear weapons too obvious not least because it was Hiroshima and also because it would have coincided with the tenth anniversary year of the dropping of the atomic bomb. Nevertheless, convincing Hiroshima to host the exhibition, albeit in 1956, was too tantalizing to ignore in terms of the potential to sway Japanese and international public opinion towards supporting the peaceful use of atomic energy. So Fotouhi set about winning the support of the Hiroshima Municipal Government, Hiroshima Prefectural Government, Hiroshima University and local newspapers. And he was successful.[19]

To make space for the exhibition, the entire contents of the museum had to be temporarily removed.[20] Despite protests by *hibakusha* and an attempt by anti-bomb activist Moritaki Ichirō who also taught ethics at Hiroshima University, to dissuade Fotouhi from removing the exhibits, the Hiroshima Municipal Government nevertheless went ahead with the plan citing financial constraints as the city could not afford 10 million yen to build a new pavilion especially for the exhibition.[21] The exhibition ran for three weeks from 27 May through to 17 June 1956 and is said to have attracted around 110,000 visitors.[22] This was testament to how Japanese could at the same time oppose nuclear weapons yet embrace civilian nuclear power, Moritaki himself would declare in Nagasaki on 10 August 1956 that

> Atomic power, which has a tendency to follow the road to destruction and extermination, must absolutely be converted to a servant for the happiness and prosperity of humankind. This is the only desire we hold as long as we live.[23]

Many cities did not have the means to show the full-scale exhibit but the USIS in Tokyo produced three sets of small *Atoms for Peace* exhibits for loan to small newspapers. As a result, it was estimated that by April

1957, over 1.5 million Japanese had seen the smaller version in the previous four months. For example, the exhibit was shown in Takamatsu on the island of Shikoku from 10–23 November 1956. During the first week from 10–15 November, 37,000 people saw the exhibit at the Mitsukoshi Department Store. A further 4800 people saw it at the Japan America Cultural Centre after it had transferred there.[24]

The exhibit provided an opportunity for local leaders and organizations to show their support for the development of atomic energy. Co-sponsors were the Shikoku Electric Co., the Kagawa Prefectural Government, the Takamatsu City Government, the *Shikoku Shimbun* newspaper and the Kagawa U.N. Association. Kagawa University and the Takamatsu P.R. Council also provided support. As stakeholders, these organizations made the most of the exhibit. At the Mitsukoshi Department Store venue, Shikoku Electric Co. incorporated the atomic energy exhibit as an integral part of a larger, million yen, electrical science exhibit commemorating the inauguration of the company's new headquarters. The governor of Kagawa prefecture provided a guided tour of the exhibition to his fellow Shikoku governors during a regional governors' conference in Takamatsu which was held at the same time. And the *Shikoku Shimbun* publicized the exhibit and carried specially written articles by Uemura Fukushichi of Kagawa University. School children and government employees were encouraged to see the exhibition.[25]

A feature at both venues was the screening of a "telemovie". At Mitsukoshi, the screen was part of the exhibit whereas at the Japan America Cultural Centre, it was strategically placed in the show window where passers-by strolling down the city's main shopping street could easily see it. Two USIS films were screened: the entertaining, fourteen-minute, colour-animated film *A is for Atom* (1952) produced originally for General Electric Co. and subsequently acquired by the USIS and *Power for Peace* (1956), a seventeen-minute colour documentary on the larger *Atoms for Peace* exhibit that had opened in Tokyo in 1955. Visitors were also shown the films in the Centre's auditorium on a larger screen during the second week of the exhibition.[26]

The exhibit toured other parts of the island from mid-November 1956 through to January 1957 and attracted a further 40,560 people. In Kochi, the exhibit was held in conjunction with the Kochi Industrial Fair from 8–14 December 1956. In Matsuyama, it was shown at the local Mitsukoshi Department Store from 12–17 January 1957. The USIS Tokyo office reported to USIA in Washington DC that the small *Atoms for Peace* exhibit was most effective when combined with a larger scale exhibit such as was the case at Mitsukoshi Department Store in Takamatsu and at Kochi. An added attraction at both of the cities were visits by the Kyoto University physicist Kimura Kiichi who gave public lectures and participated in round-table discussions. He addressed the Ehime Atomic Energy Study Society where he stated that

> It is regrettable that some Japanese have an unaccounted fear of Atomic Energy. This fear is attributable to the historical fact that the Japanese experienced the outburst and fall-out of atomic energy before they came to grips with its peaceful uses.[27]

The exhibit had encouraged the formation of Atoms for Peace Study Societies which it was expected would form the nucleus of a Shikoku-wide Atomic Energy Forum.

Nuclear Infrastructure

As a result of the Atomic Energy Basic Law that had been passed on 19 December 1955 and come into effect on 1 January 1956, much nuclear infrastructure was being put into place to make what visitors saw in the exhibits and films a reality. This began with the establishment of the Japan Atomic Energy Commission as the regulatory body for nuclear power in Japan. Its first meeting was held on 4 January 1956, chaired by Shōriki and the physicist Yukawa Hideki was among its members. On 19 May 1956, the Science and Technology Agency was launched to manage Japanese nuclear R&D and provide administrative support for the Commission through its Atomic Energy Bureau which had originally been established within the Prime Minister's Office on 1 January 1956.[28] The Japan Atomic Industrial Forum began in March 1956. The Japan Atomic Energy Research Institute (JAERI) was created as a semi-governmental organization in June 1956, followed shortly after by the Atomic Fuel Corporation in August 1956. The latter two organizations were both located at Tōkai-mura, a village in Ibaraki prefecture.

Clark D. Goodman, Assistant Director for Technical Operations, Division of Reactor Development, US Atomic Energy Commission, visited Japan from 22 March through to 8 April 1956 at the request of the US State Department. During his first two days in Tokyo, he met with USIS staff including Frances Blakemore, Exhibits Officer.[29] He spent half his time travelling throughout Japan delivering lectures on industrial nuclear power in the USA and the other half in detailed discussions with Japanese scientists and engineers, industrialists and government officials to ascertain the current state of the development of civilian nuclear power. For example, at Hiroshima University on 26 March, he met with academics, representatives of Chugoku Electric Power Company, Mitsubishi Shipyard Company, as well as staff from the Hiroshima American Cultural Centre including its Director, Abol Fotouhi.[30] Japan was familiar to Goodman as he had spent almost a year there with his family on the Fulbright Program from August 1954 through to July 1955.

In a report dated 18 April 1956, Goodman concluded that Japan was now more prosperous than in the previous two years and friendlier towards the USA compared with 1954 and 1955. Atomic energy "has now become an accepted reality to the Japanese people"[31] and a major means of international cooperation. As such, it was being used for political purposes within Japanese academia, as well as industrial and government circles. Since the Atomic Energy Basic Law had come into effect on 1 January 1956, atomic activities were being "rapidly and effectively organized."[32]

In an additional, confidential internal report dated 20 April, he provided some further background. He noted Shōriki's "meteoric rise in political power"[33] on the back of his promotion of atomic energy in Japan. He wrote that "Many well-informed, intelligent Japanese express open antipathy for Mr. Shoriki. They consider him an unscrupulous, political opportunist."[34] Despite this, Goodman was convinced that Shōriki would be "for some time the most influential person in this field in Japan."[35] In contrast to Shōriki, Goodman wrote positively about Yukawa whom he had known personally for several years. His assessment of Yukawa was a sober and fair one:

> His work on the JAEC is clearly a chore performed out of a sense of duty. He is annoyed and disgusted with the emphasis on politics. As the outstanding scientist of Japan, he enjoyed tremendous personal prestige. This is being used on the JAEC to promote the liberal interests of the Japanese scientific community. Yukawa is their spokesman and protector….I feel he is friendly toward the U.S., though at times somewhat critical of our policies.[36]

Yukawa would ultimately resign from the JAEC in March 1957, citing health reasons. Many Japanese scientists opposed the importation of reactors and were in favour of taking more time, developing a domestic R&D capability and relying more on Japanese scientists and technology.[37]

The proposed JAERI site at Tōkai-mura, Ibaraki prefecture, had just been decided after much negotiations. It was a compromise between Camp McGill, a US military base at Takeyama, Kanagawa prefecture which had originally been selected by Shōriki and the JAEC and a site at Takasaki, Gunma prefecture in central Japan that powerful politician Nakasone Yasuhiro, chairman of the Joint Diet Committee on Atomic Energy had recommended as it was his home town.[38] Now that the site had been selected, and the breaking of ground had occurred in October 1956,[39] JAERI planned to prepare for the delivery and installation of a water-boiler reactor.

Goodman noted that in the past ten months, "Japan has changed from one of considerable scepticism and some cynicism to one of remarkable awareness and receptivity toward the peaceful side of the atom."[40] In his opinion, no other country had embraced the Atoms for Peace program so favourably. This was evidenced by the JAERI site selection process where local residents had strongly supported the nomination of their respective sites as the location for Japan's first research reactors.[41] The residents seem to have been motivated by the potential economic spinoffs for their localities and concerns about radiation seem to have receded into the background. Goodman attributed this remarkable change in Japanese attitudes towards atomic energy to the "technically accurate and beautifully executed"[42] USIS *Atoms for Peace* exhibit. He had participated in the opening of the *Atoms for Peace* exhibit in Osaka on 24 March 1956 and visited a second time on 2 April.[43] Goodman acknowledged that the USA had competition in the form of the UK which was "making a strong bid to gain a share of the Japanese nuclear energy business."[44] He noted that Sir Christopher Hinton had been invited to visit Japan in May 1956.

Christopher Hinton and the Calder Hall Reactor

Sir Christopher Hinton and his wife Lillian arrived in Japan on 16 May 1956 for a twelve-day visit as guests of the *Yomiuri Shimbun*. Hinton, director of the Industrial Division of Britain's Atomic Energy Authority, revealed that the atomic power plant at Calder Hall would be completed in October. He had overseen the building of the Calder Hall reactor and not shy in stating the power plant was built for both power generation and plutonium production (for the nuclear weapons program). On his first day in Tokyo, Hinton met with Prime Minister Hatoyama Ichirō and Shōriki who was State Minister in charge of atomic energy at the time. The meeting was duly recorded by the many reporters and cameramen gathered there.[45]

The following day, he had lunch with members of the Japan Atomic Industrial Forum in Tokyo and discussed technical problems of atomic power generation.[46] On 18 May, Hinton assured Japanese gathered at a news conference that "there is no greater hazard in atomic power plants than in other industries," adding that "properly designed and properly managed plants are far safer than other well managed industries."[47] His visit was prompted by a desire to give Japan the benefit of ten years of British experience and assist Japan's own development of atomic energy for industrial use. He acknowledged that British nuclear development had begun in 1946 for defence purposes but that since 1953, its industrial use had become the first consideration. He said that electricity could be produced at very low cost in atomic plants but that the pressurized water reactors that were used in the USA cost more "than Britain can pay."[48] In making his case, Hinton said that Britain was "very anxious to consider the possibility of helping"[49] Japan through bilateral agreements.

On 19 May, Hinton gave a free public lecture on atomic power generation at Chūō University in Tokyo that was accompanied by a two-reel, colour film on the Calder Hall reactor, courtesy of the British Embassy.[50] The day after, he travelled to Osaka on 20 May to attend the inaugural meeting of the Kansai branch of the JAIF and after a busy schedule of further events including meetings with Japanese scientists, he left Japan on 28 May. Americans would later accuse Hinton of misleading the Japanese when he claimed, during his visit, that British power reactors were competitive with conventional power stations.[51] Shōriki had assumed that that would be the case in Japan, as well whereas Japanese calculations had shown that they would be at least 10 or 15 per cent higher.

If they hadn't realized before, Hinton's visit would certainly have made clear to the Japanese that he met with that although much had been made of the division between civilian and military nuclear power, in Great Britain, the former grew out of the military nuclear program. The US Atomic Energy Act of 1946 known as the McMahon Act forced Britain to follow its own path to nuclear development. The British government suspected that the real motives behind the act were commercial and that the USA was seeking to exclude Britain from the potentially lucrative market of civilian nuclear power.[52] Also, the quickest way to a nuclear weapon was through the production of plutonium in a nuclear reactor. The British eventually opted for a gas-cooled, graphite moderated, natural uranium reactor.[53]

Atoms for Peace in Hiroshima

Shortly after Goodman returned to the USA, he attended a meeting with Charles L. Dunham, Director, Division of Biology and Medicine, US Atomic Energy Commission, George S. Spiegel, State Department, and two officials from the Agreements Branch, Division of International Affairs. The meeting on 18 April 1956 was to discuss a proposal to make a medical reactor available to Japan for construction in Hiroshima. The proposal dated back a year earlier to March 1955[54] and had been the initiative of John C. Bugher who at the time was the Director of the Commission's Division of Biology and Medicine. Bugher had envisaged the USA making a financial contribution towards the reactor in the hope that it would strengthen the relationship between the Atomic Bomb Casualty Commission and the Hiroshima University medical school especially with respect to the study of radiation effects on human beings.[55]

The meeting learnt that after various discussions, the president of Hiroshima University had informed Dunham, the present Director, in a letter of 1 November 1955 that the university was planning to establish an institute for the biological and medical studies of atomic energy. He hoped that the AEC might consider assisting by providing a medical reactor. The US Department of State had advised in a memorandum of 27 March 1956 that it would support the construction of a medical reactor in Japan. In Goodman's opinion, such a request should come from the Japan Atomic Energy Commission. Dunham added that the AEC didn't have the funds and Spiegel clarified the Department of State's position by saying that its support of the proposal was premised on the understanding that the AEC would subsidize the construction of a reactor. While Dunham stated that the ABCC would be better served if a reactor was located at a Japanese university, the Hiroshima University medical school was not highly rated. What's more, laboratory facilities would first have to be developed and personnel trained. Although Bugher who was now associated with the Rockefeller Foundation had had discussions with Tsuzuki Masao, Tokyo University radiologist and Japan's representative on the UN Scientific Committee on the Effects of Radiation, in New York, it became apparent that the Foundation was not interested in financing a reactor, despite Tsuzuki's public statement otherwise.[56]

At the meeting of 18 April 1956, it was agreed that the AEC should inform the State Department that the principal advantage of a gift of a reactor would be to assure the continued access to survivors by the ABCC. But since access was currently not likely to be denied and given that Hiroshima University was at that time not scientifically capable of utilizing a medical reactor, the gift of one should not be pressed. Even if the time came when it would be advantageous to the AEC program to gift one, it should not be imposed in the absence of a specific request from the Japanese government. The AEC did not have funds for such a project and there was also the issue of reactor fuel.[57]

The following month, Takikawa Yukitoki, president of Kyoto University met with three AEC representatives: Admiral Paul Foster, Clark Goodman and Robert N. Slawson. Takikawa expressed his university's interest in procuring a pool type research reactor and was interested in any assistance that the US government might be able to provide. Foster explained that the AEC's relationship with Japan's atomic energy programme was under the Agreement for Cooperation with Japan and on a government-to-government basis. A scheme to assist countries in constructing research reactors had been announced in June 1955 but unfortunately the US contribution was limited to one reactor per country and the Japanese government had already been in contact regarding a research reactor which they wished to have US funds apply. Goodman added that the Japanese government had already contracted for a water-boiler research reactor and had plans to construct a CP-5 research reactor. Under the agreement, there would not be sufficient ^{235}U fuel available to also cover the pool type research reactor that Kyoto University was interested in building.[58]

In June 1956, Michael Sapir and Sam J. Van Hyning published *The Outlook for Nuclear Power in Japan* with the National Planning Association in Washington DC, as part of a series of reports on "the productive uses of nuclear energy." They noted that public understanding of nuclear power would be key in Japan due to what had happened at the end of the war. It acknowledged that the hazards of power reactor operation were not well understood anywhere but "the problem of understanding is of a special character for obvious reasons."[59] They realised that reassuring statements from foreign sources might not necessarily be all that helpful and that the problem had to be tackled by the Japanese themselves, notably political leaders as well scientific experts such as engineers, chemists, physicists and medical doctors.[60] The interest expressed by both Kyoto and Hiroshima Universities reflect how Japanese scientists were intent on embracing the possibilities opened up by the peaceful uses of atomic energy and looking to the AEC for funds.

Nuclear Salesmanship

Under a bilateral research agreement between Japan and the USA, a contract between JAERI and Atomics International, a division of American Aviation, Inc. had been signed on 12 March 1956 for the purchase of a 50 kW water-boiler research reactor which contained about 1.5 kg of U-235. Japan was the first country to sign such a contract for a research reactor under such arrangements.[61] The first research reactor began operation in August the following year, just before the end of the *Atoms for Peace* exhibition tour. The reactor was the first in Eastern Asia and both it and the next research reactor were products of American companies. The second American reactor was a CP-5 type research reactor with enriched uranium. But the Americans had competition.

The Americans criticized the Calder Hall type reactor that the Japanese were considering, calling it a "Model T" reactor which essentially was "the poor man's plant for countries without the resources to provide the enriched fuel required by more advanced reactor types."[62] The Japanese understood that costly fuel could potentially be the economic undoing of their civilian nuclear power program and were mindful of the terms on which fuel would be provided. The USA had insisted early on that the spent fuel, along with whatever plutonium it contained, would be removed and processed back there.[63]

Despite such criticism, when Ishikawa Ichirō, member of the Japan Atomic Energy Commission and leader of a Japanese delegation on a study mission to the UK, attended the opening of Britain's Calder Hall reactor by Queen Elizabeth II on 17 October 1956, he could not help but be impressed.[64] The excitement was heightened as official inspections of the reactor were not permitted until after the Queen had opened the facilities. Under a blue sky and with flags fluttering in the wind, the Queen, standing on a ceremonial dais, pulled a small lever which sent atomic electricity to British homes and factories.[65] A meter on the administration building roof recorded the kilowatts of electricity flowing into the commercial supply network. A photograph of the Queen inspecting the facilities graced the front pages of *The Illustrated London News*. It marked the beginning of "a new age of almost limitless possibilities for the good of mankind – if mankind so wills."[66] The Japanese were eager to share those possibilities as well, as were the representatives from thirty-eight other nations at Calder Hall, Cumberland. In an understated example of salesmanship, the Queen said, in a somewhat condescending manner that

> above all, we have something new to offer the peoples of undeveloped and less fortunate areas. It may well prove to have been among the greatest of our contributions to human welfare that we led the way in demonstrating the peaceful uses of this new source of power.[67]

Indeed, the *Economist* optimistically considered the opening of the Calder Hall power plant as marking "the formal inauguration of what will become an atomic power network stretching across Britain and ultimately across the world."[68] The technological spectacle of the "panoply of flags, flowers and presentations"[69] of the ceremony set against the backdrop of the world's first working, full-scale power reactor was overwhelming. As the Queen herself stated,

> So quickly have we learned to accept the pace of modern development that we have been in danger of losing our sense of wonder. That sense has been dramatically restored by advent of the atomic age.[70]

While the Queen drew attention to how the atomic age could potentially lead to a re-enchantment of the world in terms of how it might realise people's dreams, *The Economist* noted on a pragmatic note that the Calder Hall reactor "stands as the world's first demonstration that atomic power can be generated, not in miniature, but on the scale on which electricity is normally generated." It presented "atomic energy as a practical reality".[71] For the invited guests, seeing the real thing in contrast to models and designs at exhibitions, made a big difference. Although much was made of how the Calder Hall reactor could produce electric power quickly, cheaply and safely, the official programme for the opening did not hide the fact that the reactor was built primarily to provide more military plutonium with electricity seen as a by-product. Natural uranium was seen as being more efficient in terms of producing weapons-grade plutonium than enriched uranium.[72]

For those note able to attend the opening, the British Council sponsored the production of a nineteen-minute documentary entitled *Atomic*

Achievement (1956).[73] The film noted how the Queen had opened the Calder Hall plant in October 1956 and how it was the first full-scale atomic power station in the world. It contrasted the bright clean buildings housing uranium with their "grimy coal fired relatives".[74] The film also noted that Britain exported more isotopes than any other country in the world. The British Council reported some success in facilitating the export of British radioactive isotopes to Japan.[75]

On 17 November 1956, *The Times* reported that, thanks to having spent a month in Britain, inspecting the Calder Hall power station, visiting the Atomic Energy Research Establishment at Harwell, as well as facilities at Windscale, Springfield and elsewhere, he and his group had been able to "see the result of the well-organized activities which Britain had been doing for a long time."[76] To them, the British industrial power reactor seemed to be one of the most suitable for Japan. A second mission was sent to the UK by the newly formed Japan Atomic Power Co. that very month. They, too, wanted to see for themselves and on their return to Japan on 27 November 1956, the investigative mission's deputy head and the company's vice president Ipponmatsu Tamaki recommended to State Minister Shōriki that Japan should purchase an improved model of the British Calder Hall type reactor as its first commercial nuclear power reactor. The recommendation was made on the basis that (1) the reactor could be operated on a paying basis, (2) the natural uranium fuel could be purchased from Great Britain, and (3) the operational safety was very high.[77]

Not to be beaten, the US government lobbied the Japanese government to reconsider or at least to buy an American enriched uranium reactor as well. *The Times* of London noted the aggressive marketing of American nuclear reactor manufacturers in contrast with "the complacency and apparent lack of interest of their British counterparts."[78] The newspaper noted that Westinghouse Corporation had offered an enriched uranium Westinghouse pressurized water reactor for the giveaway price of US$35 million, on the understanding that the US Export-Import Bank would advance 80 per cent of the price, repayable over twenty-five years. Despite such favourable terms, the Japanese still preferred the Calder Hall type reactor as it was a tried and tested design that seemed to meet their "need for the earliest possible development of nuclear power."[79] They were hopeful of obtaining favourable financing from the British as well.

On 13 May 1957, the US-Japan Joint Atomic Industrial Forums Conference opened at the International Conference Hall of the Sankei Building, adjacent to the Imperial Palace in Tokyo. A delegation of 76

American representatives were joined by some 400 participants from Japan and some 70 observers from government, industry and educational institutions in the Asia-Pacific region.[80] Douglas MacArthur II, US Ambassador to Japan, spoke at a dinner for visiting American and other foreign delegates at the conference that evening. He noted that the USA had supported the formation of the International Atomic Energy Agency (IAEA) and signed atomic energy agreements with thirty-seven countries. These agreements provided an international framework for both industry and science to work together and promote peaceful uses of atomic energy and bring "more power to a world at peace."[81] He, like others at the conference entertained the hope that through the peaceful uses of atomic energy

> our lives will be made longer and healthier; our food supplies will be increased; our industries will become more productive; in fact transformed, through the availability of new and almost unlimited sources of power.[82]

This was the dream that the Japanese who had visited the *Atoms for Peace* exhibit and others throughout the world were buying into.

The *Mainichi Shimbun*, along with other Japanese newspapers, welcomed the conference. Japan was poor in energy sources and badly needed nuclear power to replace hydro-electric and thermal power generation. However, Japan lagged advanced nations such as the USA in terms of atomic industry technology and know-how and also behind in basic research in atomic energy. The *Yomiuri Shimbun* took the line that the situation called for an early purchase of power reactors. While it understood that atomic fuel supplying nations might impose restrictions on the disposal of plutonium, if these were too stringent, they would impact on Japan's sovereignty and hamper research activities. It pointed to how not only the USA and Britain but also the Soviet Union were seeking markets for their reactors and atomic fuel, so it urged the government to negotiate with that in mind.[83]

On 13 May 1957, the opening day of the conference, Prime Minister Kishi Nobusuke appeared before the House of Councillors Budget Committee. He was questioned by Socialist members of the committee who cited Kishi's recent statement that Japan could possess nuclear weapons if they were for self-defence. They suspected that Kishi would not be in a position to prevent the US from bringing nuclear weapons into the country. He denied that his government would arm the Japanese Self

Defence Forces with nuclear weapons or permit the USA to station "atomic troops" in Japan.[84]

The conference was spread over some five days with the first three days held in Tokyo, followed by a day each in Osaka and Nagoya. In the morning of the first day of the conference, Clark Goodman presented a paper entitled "The United States Power Reactor Program" on behalf of W. Kenneth Davis, Director of Reactor Development at the US Atomic Energy Commission. Davis praised American reactors which used enriched uranium in contrast to British reactors which used natural uranium.[85] This was despite a recent study by the Atomic Energy Division of American Radiator and Standard Sanitary Corporation which was sponsored and recently released by the AEC, having reached the conclusion that the British Calder Hall type reactor, if built in Japan, would have a slight initial economic advantage.[86]

The report left unmentioned the fact that enriched uranium was only available in the USA whereas natural uranium could be sourced from many nations.[87] Mori Kazuhisa, former Executive Managing Director of the JAIF, suggested that this reluctance to rely on only one nation for nuclear fuel was a key reason why a British power reactor using natural uranium was imported earlier than US light-water reactors.[88]

British newspapers responded to Davis' claims by claiming that the USA was attempting to wreck British atomic sales to Japan. *The Times* regretted how "the type of sales talk which attempts to sell by running down an opposition product comes regrettably from a representative of a United Sates Governmental organization." What's more, the newspaper claimed that "much that Mr. Davis said is, moreover, at variance with the privately expressed views of members of his own organization."[89] *The Sunday Times* was even more despondent and claimed that Japan would probably not now buy the Calder Hall type power reactor thanks to the "outspoken criticism and vigorous lobbying"[90] by US representatives at the atomic energy forum. *The Economist* accused the American "atomic salesmen in Tokyo" of attacking British atomic plants as a smokescreen to draw attention away from the fact that their own projects were running late and over budget.[91] The "unprincipled trading methods" of the Americans were also raised in the House of Commons in London. These methods were deemed "unscrupulous" especially given that Britain was the only nation that had a commercial reactor that was working.[92]

The organizers of the conference felt that it was important for the many delegates and observers, as well as members of the general public, to gain a visual understanding ("*me kara rikai saseru*") of nuclear power.[93] To this end, they held a US-Japan Atomic Industries Exhibition on the sixth floor of the Shirokiya Department Store, Nihonbashi, Tokyo. In contrast to the USIS exhibit that had been touring Japan, this exhibition consisted of models and examples of the latest equipment provided by almost sixty Japanese and US companies involved in the atomic energy industry.[94] This was not only for educational purposes. These companies were spruiking for business and the Americans appeared to have gained ground as a result of the conference. The exhibition also travelled to Osaka.

Richard L. Doan, director of the atomic energy division of Phillips Petroleum Company in Idaho, offered a slightly different take from that of his fellow Americans. He had been director of the University of Chicago's Metallurgical Laboratory that was part of the Manhattan Project that built the atomic bomb. In his eyes, it was not absolutely necessary for Japan to buy an atomic reactor from Britain or the USA. Rather, with the help of nuclear physicists in either of the countries, Japan could make up for the ten years of experience that Hinton had boasted of, in a matter of two or three years. This was more in line with the position of Japanese scientists who preferred domestic development of nuclear reactors.

Doan did, however, echo his compatriots in suggesting that Japan would be ill-advised to purchase a Calder Hall type reactor. Such a reactor would be too big for a ship, plane or submarine to carry on board. Furthermore, the British went down the Calder Hall reactor path as they lacked supplies of enriched uranium that the USA had been building up for the past dozen years. Nevertheless, if the Japanese did purchase such a reactor, he recommended that "they could take the construction plans and vastly improve the reactor on the first attempt"[95] which seems to be what the Japanese did.

In order to encourage the support and interest of young Japanese, the JAIF, in conjunction with the Tokyo Youth Council, the Science News Agency and the conference, organized an "'Atoms for Peace' and Youth" meeting at the same time and venue as the Tokyo part of the conference, 13–15 May. The young people were treated to a programme of special lectures by distinguished speakers along with films on atomic energy. On the first day, the JAERI director in charge of planning, Sagane Ryōkichi spoke on atomic energy, followed by a lecture on "American Youth and Atomic Energy" by Frederic de Hoffman, Vice President, General Dynamics Corp and General Manager, General Atomic Division. On the second day, the politician Nakasone Yasuhiro delivered a lecture on atoms for peace, followed by a lecture on the atomic age by Princeton University physicist Henry D. Smyth, President, American Physical Society and advisor to President Eisenhower when he prepared his famous atoms for peace speech. On the third and final day, Anzai Masao, Vice President, Shōwa Denkō spoke about "Atomic Energy and the Future of Japanese Industry" followed by Walter G. Whitman, Head of the Department of Chemical Engineering at MIT and former Secretary-General, UN International Conference on the Peaceful Uses of Atomic Energy, Geneva, 1955. Whitman presented a lecture on international developments relating to atomic energy.[96]

A panel of Japanese industrialists spoke at the main conference that same day and stated that they would like to build their own reactors but in the meantime, needed American help and know-how in order to do so. Kurata Chikara, chairman of the Japan Machinery Federation said that Japan's ultimate goal was domestic production of reactors but this required technical information through bilateral agreements and the training of Japanese engineers in the USA. Setoh Shoji, managing director of Tokyo Shibaura Electric Co. (also known as Toshiba) noted the need for earthquake-proof reactors. Takai Ryōtarō, president of the Tokyo Electric Power Co. (known for short as TEPCO) sounded a note of urgency, citing the need for "atomic energy to be introduced in the near future."[97]

Clark Goodman also spoke in his own right that day, noting the change in Japanese attitudes to atomic power development from when he was a Fulbright scholar and teaching at Osaka University three years prior, and now, when people almost seemed impatient to move ahead.[98] Indeed, in the aftermath of the conference, nine Japanese electric power companies banded together to form the Japan Atomic Power Co. (JAPC) with Yasukawa Daigorō as President and Ipponmatsu Tamaki as Vice President. It was envisaged that the company would enable the power utilities to jointly purchase a British Calder Hall type reactors and an American pressurized water reactor from the Yankee Atomic Electric Company. In this way, the Japanese could open the way to operating power reactors of both the enriched uranium and natural uranium types.[99]

The Japanese were in a hurry. On 27 May, the first shipment of American enriched uranium arrived on a Japan Air Lines plane in Tokyo and taken to the Japan Atomic Energy Research Institute at Tōkai-mura, Ibraki prefecture, the following day for use as fuel in an American-made 50 kW water-boiler type experimental reactor in July.[100] The equipment was purchased from North American Aviation.[101] Hubert Schenck, former Chief,

Natural Resources Section, GHQ, SCAP from 1945 to 1951, wrote the following year that "No one doubts that Japan will become a major market for power reactors."[102] American companies subsequently sold billions of dollars of equipment, technology and fuel to Japan, thereby determining what nuclear power would look like in Japan. This resulted in what has been called the world's longest standing partnership and cooperation in civilian nuclear power.[103]

Grand Exhibition of the Reconstruction of Hiroshima

The change in attitude that Goodman noted helps us understand how, in 1957, after the USIS *Atoms for Peace* exhibit tour had come to an end, parts of it were acquired by the Hiroshima Peace Memorial Museum and also by Ibaraki prefecture where the Japan Atomic Energy Research Institute was located. The announcement of presentation of the materials had been made on 28 March 1957 and it was reported in April that the USIS Tokyo office was preparing plans for the final presentation of materials. This included all the USIS's own equipment but inquiries were made to the USIA in Washington DC as to whether any of the borrowed equipment could be purchased and donated for permanent showing. USIS Tokyo thought that three items were worth paying the asking price: the model of a portable power reactor made by American Locomotive; a fume hood produced by P.M. Lennard, and a superscalar counter produced by Tracer-lab that could measure radioactivity.

Other items such as a swimming pool reactor by the AVCO Corporation were deemed as over-priced but if obtainable at a large discount, was also on the wish-list.[104] "Swimming pool" reactors have been a common form of nuclear reactor used for research purposes. A reactor core of fuel elements and control rods are suspended near the bottom of a large, deep, open tank or pool of water. Enriched uranium is used as fuel and light water as moderator and coolant. The core is shielded by maintaining a sufficient depth of water over the core and by the solid concrete walls of the tank. Concrete could be cut away and aluminium windows inserted to enable experiments to be conducted.

Mutō Ichiyo who was at the time a member of staff in charge of the International Section, Japan Council against Atomic and Hydrogen Bombs (known in Japanese for short as Gensuikyō), visited the Hiroshima

Peace Memorial Museum in 1957 and recalls how after viewing the very moving and dimly-lit museum display documenting the effects of the atomic bombing, he entered a bright room, the Atoms for Peace annex. It was here where the highlights of the USIS exhibit had been deposited: "magic hands" holding nuclear materials, a model of an atomic airplane, atomic-powered ships and trains, and a model nuclear power plant. He noted how only a thin door separated what appeared to be two completely different worlds: hell and paradise.[105]

Both worlds would be on display as part of the *Hiroshima Fukkō Daihakurankai*, (*Grand Exhibition of the Reconstruction of Hiroshima*) which ran from 1 April through to 20 May 1958. The exhibition consisted of some thirty-seven pavilions spread over three venues: Peace Park, Heiwa Ōdōri (Peace Boulevard) and Hiroshima Castle. There were eighteen pavilions in the main venue of Peace Park. These tended to be scientific in nature and included the Peace Memorial Museum and its annexe which became the pavilion for the Peaceful Use of Atomic Energy.[106] Other pavilions at Peace Park included an Electrical Science pavilion which explained to visitors how electricity was generated by nuclear reactors. The Museum and its annexe were second only to the Space Exploration pavilion in popularity. That pavilion was among thirteen pavilions and attractions located in the second venue along Peace Boulevard.[107] It is estimated that 920,7261 tickets were sold in total to the *Grand Exhibition*.[108]

Japan International Trade Fair, Osaka

Meanwhile, not far away, the USSR participated for the first time in a post-war Japanese trade fair. The Japan International Trade Fair, which had alternated each year between Tokyo and Osaka, was held in Osaka that year from 12–27 April. The Soviets showcased their know-how in an informative atomic energy display in their pavilion, a photograph of which graced the front page of the *Genshiryoku Sangyō Shimbun* (*Atomic Industry Newspaper*).[109] The USIS, Tokyo office and US Embassy, Tokyo were dismayed to find that in 1958, 66 per cent of the 500 visitors interviewed had chosen the Soviet pavilion as the most interesting, in contrast to the 11 per cent who had chosen the American pavilion. Thirty-nine per cent of respondents had chosen the USSR as being the most advanced in science and technology and 20 per cent chose the USA. These figures reflect the impact of the launching of Sputnik 1 just six months before on 4 October 1957 and Sputnik 2 on 3 November 1957 with the Soviet space dog Laika on board. Models of Sputnik 1 and Sputnik 2 dominated the Soviet exhibit at the Osaka trade fair and were very popular amongst visitors.[110]

Expo '58

At the same time as the *Grand Exhibition of the Reconstruction of Hiroshima* was seeking to introduce the Japanese public to the atomic age by once again showing the "magic hands" that could handle nuclear materials remotely, "hands" were also a feature of the Japanese pavilion at Expo '58, the world's fair held in Brussels from 17 April through to 19 October 1958. Described as the world's first Atomic Fair, its striking centrepiece was the 102-metre high Atomium building which consisted of nine, stainless-steel spheres, mimicking atoms in an elementary crystal of iron. Each sphere served as exhibit halls for displays on atomic energy.[111] The fair provided Japan with an opportunity to project its narrative of defeat, peace and reconstruction to an international audience.

A few years earlier, *The Washington Post* had reported on the race between Belgium and to Japan to have the first atomic power plant under the auspices of President Eisenhower's Atoms for Peace plan. While Japan was the first nation to have the atomic bomb used against it in wartime, Belgium had the distinction of having African mines which had made the "largest single contribution of uranium to the United States atomic program."[112] Expo '58 served to underline this. Japan's participation showed the world that it had rejoined the family of nations and wished to share in the benefits of the atomic age.

The American pavilion and its exhibit entitled "The Face of America" were located directly across from the Soviet pavilion. One month into the fair, the *Yomiuri Shimbun* carried an article on this "atomic age" fair which compared the US and Soviet pavilions and included photographs of both.[113] As with the Japan International Trade Fair at Osaka, it was a challenge for the USA to match the Soviet pavilion with its models of Sputnik.[114] Any Japanese visitors would have found part of the display on atomic energy in the US pavilion familiar, especially the "magic hands" that enabled visitors to play with radioactive substances without fear.[115] The emphasis on the American way of life would also have been reminiscent of the 1950 America Fair.

The Japan pavilion, designed by the architect Maekawa Kunio,[116] sought to show the world that the Japanese had put World War II behind them. Photographs by Watanabe Yoshio provided visitors with images showing how far they had come.[117] Their efforts were symbolized in the very theme of the pavilion: "The Japanese Hand and the Machine." Told as a story in three parts, the first section which focused on history, traced how the Japanese hand and ingenuity had crafted objects throughout time including pre-modern artefacts such as ancient funerary clay sculptures known as *haniwa*, to the reproduction of a Buddha head from Kōfukuji temple in Nara.

A photograph of an open hand was accompanied by a didactic panel which explained how "the Japanese Hand, which has worked and created throughout so many ages" had woven "the thread of its destiny."[118] In the background, traditional music was played to convey a sense of Japan's rich cultural legacy. The section ended abruptly with a large photograph of Hiroshima shown soon after the dropping of the atomic bomb. All visitors to the pavilion would walk under this image of destruction and come out the other side to see an image of a group of miners and a large, red truck, representing industrialized Japan.

This ushered visitors into the next section on industry where they would see a photograph of a clenched fist, accompanied by the words "after the misery and destruction of war, the Japanese Hand again begins to toll, unweariedly, and with renewed cheerfulness."[119] The section showcased advanced, scientific instruments including electron microscopes, Geiger counters, lenses, and cameras, proof that the Japanese hand was dexterous and suited to precision work. Visitors were reminded that "The Japanese Hand is one not only of strength and power but of finesse, a hand of thought and intelligence"[120] by both text and the image of an index finger of a right hand plucking at a string or wire.

The third and final section was devoted to everyday life and the role that traditional handicrafts such as textiles, ceramics and basketry played. A didactic panel conveyed how "in this Machine Age, the Japanese Hand continues to weave its destined part of that thread which leads humanity 'tomorrow to fresh woods and pastures new'."[121] In this way, Japanese hands signified the Japanese people, their national character, dexterity and work ethic. They had rebuilt the nation after the war and embraced modern science and technology while maintaining cultural traditions which continued to inform everyday life. Importantly, Expo '58 placed this in the context of the atomic age and the cold war rivalry between the USA and the USSR.

The US-Soviet rivalry could also be seen at the Second United Nations International Exhibition on the Peaceful Uses of Atomic Energy in

Geneva, 1–13 September and the associated conference. In relation to the exhibition which was being held at the Palais des Nations in Geneva, the USIS released to the *Japan Times* newspaper a selection of photographs and illustrations providing visual evidence that supported the claim that by 1964, just when the British Calder Hall reactor at Tōkai-mura was expected to be completed, the USA would have in place atomic power plants producing 1.3 million kilowatts of electricity. They would also have airplanes and ships with atomic reactors of equal capacity on board for propulsion. The images and associated article took up an entire page of the newspaper. The article did concede that "because the nation [USA] has ample coal and oil, its atomic-produced electricity will not be competitive in cost for many years." Despite this, the US government and industry were committed to spending billions of dollars to develop economical atomic power reactors. Over the next six years, it was envisaged that the present eight civilian nuclear power plants would increase to twenty-three. In contrast to Great Britain, the USA would endeavour to find "not one but several types of power reactors capable of economic operation in fuel-short nations in the near future."[122]

Meanwhile in Geneva, a fifty-seven-person delegation of Japanese officials, industrialists and scientists attended the exhibition and viewed the latest nuclear equipment assembled by the bigger nuclear nations. They were among the 5000 delegates and observers from 66 nations in attendance. Ishikawa Ichirō (JAEC) was there as was the theoretical physicist Yukawa Hideki. The Japanese mounted a small display showing achievements in industry and agriculture, including a chart showing how the use of radioisotopes had enabled them to breed useful mutations of rice and increased crop yield. Another chart illustrated Japanese uranium mining.[123] This was an extremely modest exhibit in contrast to the US exhibit. Yukawa remarked that

> the Americans brought the real equipment to Geneva, instead of just drawings and papers. We realize that this is a very expensive affair. Their exhibit is a very neat and compact show of the latest developments, particularly in the field of fusion.[124]

As the *Japan Times* noted, the technical exhibits were meant to be studied by scientists in conjunction with the papers presented at the conference.[125] Yukawa remarked that the US exhibit revealed information and machinery that is not known in Japan, despite a constant flow of information to Japan from the USA. In the field of nuclear fusion, Japan was far behind the USA, Britain and the Soviet Union. He acknowledged that the Japanese had benefited greatly from the last Geneva conference in 1955 and was confident that it would also be the same at the second conference.[126] Indeed, the information on thermonuclear reactions, fusion, that Yukawa referred to, had been newly declassified by the USA and Britain in the hope that this would lead to a freer flow of information and research cooperation. Nuclear power through controlled fusion was not likely to be realized for many years.

In its editorial on 2 September 1958, the *Japan Times* suggested that in light of the Geneva conference and exhibition, "no nation could afford to stand aloof from this great scientific movement." Japan had made modest but steady progress in the peaceful uses of atomic energy and there had been Japanese representation at the conference and an interesting exhibit, but "it is obvious that the painstaking labors of our scientists must have the support of the nation. No opportunity should be lost in encouraging the study of nuclear science in this country."[127] The eminent British scientist Sir John Cockcroft delivered a lecture at Geneva on the trends of the development of the peaceful uses of atomic energy[128] and soon after travelled to Japan where he met with Japanese scientists and business men to allay their concerns about the improved Calder Hall reactor that the Japan Atomic Power Co. was planning to purchase.

Cockcroft arrived on 17 November and was given a tour of what the Japanese had accomplished at Tōkai-mura. Together with Sagane Ryōkichi who had become deputy director-general at JAERI and Ipponmatsu Tamaki, JAPC vice president, Cockcroft inspected the water-boiler type reactor that was in operation and viewed the CP-5 reactor that was being installed.[129] On 19 November, Cockcroft delivered a lecture that was jointly hosted by the JAIF and the Science Council of Japan.[130] In this way, he was able to meet with both industrialists and scientists at the same forum and allay Japanese concerns about a number of things including the earthquake proofness of the reactor, the build-up of Wigner energy that had caused the recent accident at Windscale, and the "positive temperature coefficient" that he said could be controlled. Also of apparent concern were reports that the UK Atomic Energy Authority had modified the design of reactors so that the plutonium produced might be suitable for military purposes in order to compensate for the high cost of producing electricity. Another query was in relation to the economics of nuclear power production. Cockcroft's visit served to restore Japanese confidence in the choice of the Calder Hall reactor but it was predicted that the signing of a formal contract would have to await until the following year.[131]

Conclusion

By the end of 1958, thanks to efforts by the USA and Great Britain to promote their reactors, there was the perception in Japan that commercial nuclear power generation was "just around the corner" although power through nuclear fusion seemed to be a distant prospect.[132] American supremacy in science and technology since the end of the war was challenged by the British Calder Hall reactor, Soviet efforts in atomic energy and the launch of Sputnik 1 and 2. Japan signed agreements with both the USA and Great Britain to facilitate the importation of atomic power plants but ultimately chose to place an order for an improved Calder Hall type reactor with the British General Electric Co. on 22 December 1959 which was expected to be completed in early 1964 at Tōkai-mura.[133] The construction was not without problems and Japan subsequently opted for American light-water reactors which they bought directly from General Electric and Westinghouse and then built themselves under licence.[134] Japan's dreams of nuclear power became oriented towards the USA against the backdrop of cold war tensions.

Notes

1. Koichirō Kurihara, "Atom Boss Likely to Become Man of the Year 1956," *Nippon Times*, 1 Jan. 1956, 11.
2. "Genshiryoku heiwa riyō no yume" ("The Dream of Atoms for Peace"), *Yomiuri Shimbun*, 1 Jan. 1956, 9.
3. "Genshiryoku heiwa riyō no yume."
4. "Genshiryoku heiwa riyō no yume."
5. "Genshiryoku heiwa riyō no yume."
6. "Josei to genshiryoku" ("Women and Atomic Energy"), *Yomiuri Shimbun*, 1 Jan. 1956, 9.
7. Kanō, Mikiyo, "'Genshiryoku no heiwa riyō' to kindai kazoku" ("'Atoms for Peace' and the Modern Family"), *Jendaa shigaku* (*The History of Gender*), no. 11 (2015): 5–19, esp. 19.
8. For details see George O. Totten and Tamio Kawakami, "Gensuikyō and the Peace Movement in Japan," *Asian Survey* 4, no. 5 (May 1964): 833–841; Wesley Sasaki-Uemura, *Organizing the Spontaneous: Citizen Protest in Postwar Japan* (Honolulu: University of Hawaii Press, 2001),

122–124; Joseph Orr, *The Victim as Hero: Ideologies of Peace and National Identity in Postwar Japan* (Honolulu: University of Hawaii Press, 2001), 49–52.

9. Muto Ichiyo, "The Buildup of a Nuclear Armament Capability and the Postwar Statehood of Japan: Fukushima and the Genealogy of Nuclear Bombs and Power Plants," *Inter-Asia Cultural Studies* 14, no. 2 (2013): 171–212, esp. 173–174.
10. Aichi Prefectural Library, "Kaikan kara 50-nen" ("50 Years Since Opening"), accessed 9 Dec., 2018, https://websv.aichi-pref-library.jp/event/aichitoshokan_chirashi.pdf
11. Foster Hailey, "U.S. Top Exhibitor at Japanese Fair," *New York Times*, 5 Apr. 1956, 41.
12. "Osaka Fair Opens under Heavy Rain," *New York Times*, 9 Apr. 1956, 35.
13. David E. Nye, *American Technological Sublime* (Cambridge, Mass.: MIT Press, 1994), 216.
14. Hailey, "U.S. Top Exhibitor at Japanese Fair."
15. Foster Hailey, "Japanese Found in Dark About Us," *New York Times*, 16 Apr. 1956, 36.
16. US Government, "Participation of the Sino-Soviet Bloc in International Trade Fairs and Exhibits," Intelligence Memorandum, IM-430, Central Intelligence Agency, Office of Research and Reports, 1 May, 1956.
17. "Osaka Fair Opens under Heavy Rain."
18. "1,000,000 in Hiroshima Visit U.S. Atom Exhibit," *Nippon Times*, 2 Jun. 1956, 3.
19. Ryuichi Kanari, "U.S. Used Hiroshima to Bolster Support for Nuclear Power," *Asahi Shimbun*, 25 Jul. 2011, *Asahi Shimbun* database. See also Ran Zwigenberg, *Hiroshima: The Origins of Global Memory Culture* (Cambridge: Cambridge University Press, 2014), 114–115.
20. See NHK (Japan Broadcasting Corporation), "Hiroshima: Bakushinchi Genshiryoku Heiwa Riyō Hakurankai" ("Hiroshima: Ground Zero, Atoms for Peace Exhibition"), *ETV tokushū* (*E Television Special Report*), 2014, accessed 29 Jan. 2020, http://www.nhk.or.jp/etv21c/file/2014/1018.html
21. Moritaki Ichirō, *Kaku zettai hitei e no ayumi* (*Steps toward Absolute Negation of the Nuclear* (Hiroshima: Keisuisha, 1994). Cited in Muto, "The Buildup of a Nuclear Armament Capability and the Postwar Statehood of Japan," 176.
22. Kanari, "U.S. Used Hiroshima to Bolster Support for Nuclear Power."
23. See Hidankyō (Japan Confederation of Atomic and Hydrogen Bomb Sufferers Organizations), "Message to the World," 10 Aug. 1956, accessed 10 Dec., 2018, http://www.ne.jp/asahi/hidankyo/nihon/rn_page/english/message.html. Cited in Zwigenberg, *Hiroshima*, 127.

24. Foreign Service Despatch from USIS Tokyo to USIA Washington, "Small Atoms for Peace Exhibit," 18 April, 1957, Japan – Atomic Energy 1957, RG 469, Box 32, Records of U.S. Foreign Assistance Agencies, 1942–1963, U.S. National Archives and Records Administration, College Park, MD.
25. Foreign Service Despatch, "Small Atoms for Peace Exhibit."
26. Foreign Service Despatch, "Small Atoms for Peace Exhibit." See also Tsuchiya Yuka, "The Atoms for Peace USIS Films: Spreading the Gospel of the 'Blessing' of Atomic Energy in the Early Cold War Era," *International Journal of Korean History* 19, no. 2 (Aug. 2014), 107–135, esp. 120–121.
27. Foreign Service Despatch, "Small Atoms for Peace Exhibit;" Tsuchiya, "The Atoms for Peace USIS Films," 120–121.
28. "Activities of Japan Atomic Energy Commission in 1956," *Atoms in Japan* 1, no. 1 (May 1957): 5–7; Morris Low, Shigeru Nakayama and Hitoshi Yoshioka, *Science, Technology and Society in Contemporary Japan* (Cambridge: Cambridge University Press, 1999), 74.
29. US Atomic Energy Commission, Memorandum from Clark D. Goodman to John A. Hall through W. Kenneth Davis, "Nuclear Energy Developments in Japan", 18 Apr. 1956, Appendix B, 1, RG 59, General Records of the Department of State, Office of the Secretary, Special Assistant to Secretary of State for Atomic Energy and Outer Space, General Records Relating to Atomic Energy Matters, 1948–1962, NND 949670, Box 425, US National Archives and Records Administration, College Park, MD.
30. Goodman to Hall, "Nuclear Energy Developments in Japan", 18 Apr. 1956, Appendix B, 2.
31. Goodman to Hall, "Nuclear Energy Developments in Japan", 18 Apr. 1956, 1.
32. Goodman to Hall, "Nuclear Energy Developments in Japan", 18 Apr. 1956, 1.
33. US Atomic Energy Commission, Memorandum from Clark D. Goodman to John A. Hall through W. Kenneth Davis, "Political Notes from Visit to Japan", 20 Apr, 1956, 1, RG 59, General Records of the Department of State, Office of the Secretary, Special Assistant to Secretary of State for Atomic Energy and Outer Space, General Records Relating to Atomic Energy Matters, 1948–1962, NND 949670, Box 425, US National Archives and Records Administration, College Park, MD.
34. Goodman to Hall, "Political Notes from Visit to Japan", 2.
35. Goodman to Hall, "Political Notes from Visit to Japan", 2.
36. Goodman to Hall, "Political Notes from Visit to Japan", 3–4.
37. "A Policy on Atomic Power," editorial, *Japan Times*, 3 Dec. 1956, 8.

38. Goodman to Hall, "Political Notes from Visit to Japan", 2.
39. "Calder Hall Opening," editorial, *Japan Times*, 19 Oct. 1956, 8.
40. Goodman to Hall, "Political Notes from Visit to Japan", 5.
41. Goodman to Hall, "Political Notes from Visit to Japan", 6.
42. Goodman to Hall, "Political Notes from Visit to Japan", 6.
43. Goodman to Hall, "Nuclear Energy Developments in Japan", 18 Apr. 1956, Appendix A.
44. Goodman to Hall, "Political Notes from Visit to Japan", 6.
45. "A-Expert Arrives," *Nippon Times*, 17 May, 1956, 2.
46. "A-Expert Arrives."
47. Tak Ishii, "Hinton Says A-Reactors Safer Than Other Plants," *Nippon Times*, 19 May, 1956, 3.
48. Ishii, "Hinton Says A-Reactors Safer Than Other Plants."
49. Ishii, "Hinton Says A-Reactors Safer Than Other Plants."
50. "Genshiryoku hatsuden dai-kōenkai" ("Big Conference on Nuclear Power Generation"), *Yomiuri Shimbun*, 14 May, 1956, 7.
51. "Japanese Battle over Atomic Reactors," *Times*, 29 May, 1957, 9.
52. Stephen Twigge, "The Atomic Marshall Plan: Atoms for Peace, British Diplomacy and Civil Nuclear Power," *Cold War History* 16, no. 2 (2016): 1–18, esp. 5.
53. Roger Williams, "British Nuclear Power Policies," in, *Energy Economics in Britain*, ed. Paul Tempest (London: Graham and Trotman, 1983), 35–58, esp. 35–36.
54. See M. Susan Lindee, *Suffering Made Real: American Science and the Survivors at Hiroshima* (Chicago: University of Chicago Press, 1994), note 57, 160.
55. Robert N. Slawson, AEC, "Memorandum to the Files: Meeting on Japan – Medical Reactor Proposal," 2 May, 1956, RG 59, General Records of the Department of State, Office of the Secretary, Special Assistant to Secretary of State for Atomic Energy and Outer Space, General Records Relating to Atomic Energy Matters, 1948–1962, NND 949670, Box 425, US National Archives and Records Administration, College Park, MD.
56. Slawson, "Memorandum to the Files," 2 May, 1956.
57. Slawson, "Memorandum to the Files," 2 May, 1956.
58. Robert N. Slawson, AEC, "Memorandum of Conversation: Meeting with Dr. Yukitoki Takikawa, President of Kyoto University," 22 May, 1956, RG 59, General Records of the Department of State, Office of the Secretary, Special Assistant to Secretary of State for Atomic Energy and Outer Space, General Records Relating to Atomic Energy Matters, 1948–1962, NND 949670, Box 425, US National Archives and Records Administration, College Park, MD.

59. Michael Sapir and Sam J. Van Hyning, *The Outlook for Nuclear Power in Japan*, Reports on the Productive Uses of Nuclear Energy (Washington DC: National Planning Association, Jun. 1956), esp. 112.
60. Sapir and Van Hyning, *The Outlook for Nuclear Power in Japan*, 112.
61. Goodman to Hall, "Nuclear Energy Developments in Japan", 18 Apr. 1956, 7.
62. "Japan Backs Harwell," *Economist*, 16 Jun. 1956, 1120.
63. "Japan Backs Harwell."
64. "A-Power Program Not Yet Crystallized," *Atoms in Japan* 1, no. 1 (May 1957): 2–4, esp. 2.
65. Sir Christopher Hinton, "Atomic Power in Britain," *Scientific American* 198, no. 3 (Mar. 1958): 29–35.
66. "October 17, 1956 – The Day Which Saw the Beginning of a New Age," *The Illustrated London News*, 27 Oct. 1956, 691–693.
67. "October 17, 1956 – The Day Which Saw the Beginning of a New Age."
68. "Atomic Power Arrives," *Economist*, 20 Oct. 1956, 257.
69. C.N. Hill, *An Atomic Empire: A Technical History of the Rise and Fall of the British Atomic Energy Programme* (Singapore: World Scientific Publishing Company, 2013), 179.
70. Hill, *An Atomic Empire*, 179.
71. "Atomic Power Arrives," *Economist*, 20 Oct. 1956, 257.
72. John Krige, "The Peaceful Atom as Political Weapon: Euratom and American Foreign Policy in the Late 1950s," *Historical Studies in the Natural Sciences* 38, no. 1 (Winter 2008): 5–44, esp. 17. See also "Our Nuclear Power Stations," *Guardian*, 2 Dec. 1964, 18.
73. "Atomic Achievement: Transcript," UK National Archives, http://www.nationalarchives.gov.uk/films/1951to1964/filmpage_atomic.htm
74. "Atomic Achievement: Transcript."
75. Christopher Aldous, "Masking or Marking Britain's Decline? The British Council and Cultural Diplomacy in Japan, 1952–1970," in *The History of Anglo-Japanese Relations, 1600–2000: Vol. 5, Social and Cultural Perspectives*, eds. Gordon Daniels and Chushichi Tsuzuki (Basingstoke, UK: Palgrave Macmillan, 2002), 330–351, esp. 339.
76. "Japanese Interest in Atom Station," *Times*, 17 Nov. 1956, 6.
77. "A-Power Mission Head Suggests U.K. Reactor," *Japan Times*, 28 Nov. 1956, 3.
78. "Power Reactor Salesmanship," *Times*, 1 Apr. 1957, 9.
79. "Power Reactor Salesmanship," *Times*, 1 Apr. 1957, 9.
80. "U.S. Out to Win Reactor Race," *Times*, 14 May, 1957, 8.
81. Douglas MacArthur II, "Atomic Forum," *Japan Times*, 14 May, 1957, 8.
82. "MacArthur Calls for Atom Energy Cooperation," *Japan Times*, 14 May, 1957, 1.

83. "Press Comments," *Japan Times*, 14 May, 1957, 8.
84. "Kishi Denies Gov't Seeking A-Weapons," *Japan Times*, 15 May, 1957, 1.
85. John W. Finney, "British Reactor Given Cost Edge," *New York Times*, 19 May, 1957, 7. See also "Bei dōryoku ro, Ei ni masaru" ("US Power Reactor Superior to British"), *Asahi Shimbun*, evening edition, 13 May, 1957, 1; "Sekkyokutekna 'makikaeshi saku'" ("Aggressive 'Rollback Policy'"), *Asahi Shimbun*, 14 May, 1957, 1.
86. A. Puishes, D.P. Herron, D.R. Mash and J.W. Webster, *Comparison of Calder Hall and PWR Reactor Types*, prepared for Division of Reactor Development, U.S. Atomic Energy Commission (Redwood City, Calif.: Atomic Energy Division, American Radiator and Standard Sanitary Corporation, 1 Mar. 1957). See the Summary, 3.
87. Finney, "British Reactor Given Cost Edge;" "Bei dōryoku ro, Ei ni masaru."
88. Kazuhisa Mori, "Atomic Bombing and Nuclear Energy Development in Japan," Japan Atomic Industrial Forum, 1995, typescript, Yukawa Hall Archival Library, Kyoto University, accessed 11 Dec. 2018, https://www2.yukawa.kyoto-u.ac.jp/~yhal.oj/Mori/Mori1Q/Reports02/IMG_0025.pdf
89. "U.S. Out to Win Reactor Race." The content of the *Times* article was report on in Japan in "Davis Criticized," *Japan Times*, 15 May, 1957, 2.
90. Our Own Representative, "U.S. Ousts British Atom Plant," *Sunday Times*, 19 May, 1957, 17.
91. "Atomic Gamesmanship," Business Notes, *Economist*, 1 Jun. 1957, 815.
92. Our Parliamentary Correspondent, "U.S. Reactor Comments Deplored," *Financial Times*, 29 May, 1957, 9.
93. "Nichibei genshiryoku sangyō tenrankai" ("US Japan Atomic Industries Exhibition"), *Genshiryoku Sangyō Shimbun* (*Atomic Industry Newspaper*), 25 Feb. 1957, 1.
94. "Joint Conference Opens," *Atoms in Japan*, Japan Atomic Industrial Forum, 1, no. 1 (May 1957), 23.
95. Dick Horning, "Japan's Own Reactor," *Japan Times*, 14 May, 1957, 8.
96. "Genshiryoku heiwa riyō to seishōnen no kai" ("Meeting on 'Atoms for Peace' and Youth"), *Genshiryoku Sangyō Shimbun* (*Atomic Industry Newspaper*), 15 Apr. 1957, 3.
97. Associated Press, "Industrialists Call for U.S. Help in Building Atomic Reactors Here," *Japan Times*, 16 May, 1957, 1.
98. Kiyoaki Murata, "Nuclear Power in Japan's Future," *Japan Times*, 18 May, 1957, 8.
99. "Japan to Buy British and U.S. Reactors," *Times*, 24 May, 1957, 8; "Japanese Battle over Atomic Reactors."
100. "First Uranium Fuel Is Flown Into Japan," *Japan Times*, 28 May, 1957, 1.

101. Associated Press, "GE Proposes to Construct Atomic Plant for Japan," *Japan Times*, 28 May, 1957, 1.
102. Hubert G. Schenck, "Impact of Science in East Asia," *Bulletin of the Atomic Scientists* 14, no. 7 (Sept. 1958): 273–275, esp. 274.
103. Japan also contributed funds to US nuclear R&D programs in the 1960s and paid licence fees to the US Atomic Energy Commission and its successor bodies for the provision of nuclear fuel. See Phyllis Genther Yoshida, *U.S.-Japan Nuclear Cooperation: The Significance of July 2018* (Washington, DC: Sasakawa Peace Foundation USA, 2018), accessed 9 Dec. 2018, https://spfusa.org/research/u-s-japan-nuclear-cooperation-the-significance-of-july-2018/
104. Foreign Service Despatch from USIS Tokyo to USIA Washington, "ICS: Disposition of Atoms for Peace Exhibit," 12 Apr. 1957, Japan – Atomic Energy 1957, RG 469, Box 32, Records of U.S. Foreign Assistance Agencies, 1942–1963, U.S. National Archives and Records Administration, College Park.
105. Muto, "The Buildup of a Nuclear Armament Capability and the Postwar Statehood of Japan," 174.
106. For an account of a visit to the pavilion see Hiroshoma Fukkō Daihakurankaishi Henshū Iinkai, *Hiroshima Fukkō Daihakurankaishi* (*Report of the Grand Exhibition of the Reconstruction of Hiroshima*) (Hiroshima: Hiroshima City Hall, 1959), 98–102.
107. Yuki Tanaka, "'The Peaceful Use of Nuclear Energy' and Hiroshima" in Yuki Tanaka and Peter Kuznick, "Japan, the Atomic Bomb, and the 'Peaceful Uses of Nuclear Power,'" *Asia-Pacific Journal: Japan Focus* 9, 18, no. 1 (2 May, 2011), accessed 13 Dec. 2018, https://apjjf.org/2011/9/18/Yuki-Tanaka/3521/article.html
108. Hiroshima-shi Sōmukyoku Sōmuka, *Shisei yōran* (Hiroshima City General Affairs Bureau, General Affairs Division), *Shisei yōran* (*Municipal Handbook*) (Hiroshima: Hiroshima City Hall, 1959), 64; Tanaka, "'The Peaceful Use of Nuclear Energy'"; Hiroshima Peace Memorial Museum, "Hiroshima no fukkō" ("Reconstruction of Hiroshima"), accessed 13 Dec. 2018, http://www.pcf.city.hiroshima.jp/Peace/J/pHiroshima4_3.html
109. "Ōsaka de kaisai shita kokusai mihon'ichi no Soren genshiryoku kan" ("The Soviet Atomic Energy Pavilion at the International Trade Fair held in Osaka," *Genshiryoku Sangyō Shimbun* (*Atomic Industry Newspaper*), 15 Apr. 1958, 1.
110. Wendell D. Baker, Research Office, USIS, US Embassy, Tokyo, "Brief Summary of Methodology and Major Findings, Osaka Trade Fair Study, April 17–26, 1958," May 1958, RG306, Records of the USIA, Office of Research, Country Project Files, Japan 1958, Box 64, ARC ID 1065787,

US National Archives and Records Administration, College Park, MD; "1958-nen Nihon Kokusai Mihon'ichi kara" ("Inspecting Round the Japan International Trade Fair, Osaka"), 1958, technical paper, National Institute of Advanced Industrial Science and Technology, Tohoku Centre, accessed 29 Dec. 2018, https://unit.aist.go.jp/tohoku/techpaper/pdf/3888.pdf

111. Michelle Jeandron, "The Atomium Marks a Half Century," *Physics World* (April 2008), 7; Stephen Petersen, "Explosive Propositions: Artists React to the Atomic Age," *Science in Context* 17, no. 4 (Dec. 2004): 579–609, esp. 599–600.
112. "Belgium and Japan Seek 1st 'A-for-Peace' Power," *Washington Post*, 15 Feb. 1955, 5.
113. Nakajima (Special Correspondent), "Genshiryoku jidai no Brusseru kokusai hakurankai" ("The Atomic Age Brussels International Exposition"), *Yomiuri Shimbun*, 11 May, 1958, evening edition, 2.
114. Robert H. Haddow, *Pavilions of Plenty: Exhibiting American Culture Abroad in the 1950s* (Washington DC: Smithsonian Institution Press, 1997), 105–106.
115. Robert Hamilton Haddow, "Material Culture and the Cold War: International Trade Fairs and the American Pavilion at the 1958 Brussels World's Fair," (PhD dissertation, 2 vols., University of Minnesota, 1994), Vol. 1, 130–131.
116. Jonathan M. Reynolds, *Maekawa Kunio and the Emergence of Japanese Modernist Architecture* (Berkeley: University of California Press, 2001), esp. 212–13.
117. Ueno (Special Correspondent), "Brusseru bankokuhaku hiraku" ("The Brussels Expo Opens"), *Yomiuri Shimbun*, 18 Apr. 1958, 3.
118. "Burasseru Bankoku Haku Nihon sanka keikaku" ("Plans for Japanese Participation in the Brussels Universal and International Exhibition"), *Kōgei nyūsu* (*Industrial Art News*), 26, no. 3 (Mar.-Apr. 1958), 2–31, esp. 10.
119. "Burasseru Bankoku Haku Nihon sanka keikaku," esp. 12.
120. "Burasseru Bankoku Haku Nihon sanka keikaku," esp. 14.
121. "Burasseru Bankoku Haku Nihon sanka keikaku," esp. 16.
122. "Atoms for Peace," *Japan Times*, 1 Sept. 1958, 5.
123. Kyodo-Reuter, "Japan May Reveal New Information at H-Parley," *Japan Times*, 1 Sept. 1958, 1.
124. Associated Press, "U.S. Geneva Exhibit Reveals New Data," *Japan Times*, 2 Sept. 1958, 1.
125. "Atomic Conference," editorial, *Japan Times*, 2 Sept. 1958, 8.
126. Associated Press, "U.S. Geneva Exhibit Reveals New Data."
127. "Atomic Conference."

128. Sir John Cockcroft and Peter Collins, "Cheaper Nuclear Power in Five Years," *Sunday Times*, 31 Aug. 1958, 12.
129. "Cockcroft Visits Tokai A-Facilities," *Japan Times*, 22 Nov. 1958, 6.
130. "'Eikoku no genshiryoku kaihatsu keikaku' o setsumei" ("An Explanation of British Plans for the Development of Atomic Energy"), *Genshiryoku Sangyō Shimbun*, 25 Nov. 1958, 2.
131. Our Tokyo Correspondent, "Sir J. Cockcroft Allays Tokyo Fears about Reactors," *Times*, 27 Nov., 1958, 9.
132. Watanabe, Seiki, "The Road to Atomic Power," *Japan Quarterly* 5, no. 4 (1 Oct. 1958): 418–425, esp. 421.
133. Our Own Correspondent, "British Reactor for Japan," *Times*, 23 Dec. 1959, 6; "Japanese A-Power Contract Signed," *Financial Times*, 23 Dec. 1959, 7.
134. John Foster, "The Development of Nuclear Power," in *Special Symposium: 50 Years of Nuclear Fission in Review*, ed. Malcolm Harvey, Ottawa, 5 Jun. 1989, Proceedings of the 29th Annual Conference, Canadian Nuclear Society, Vol. 1, accessed 19 Dec. 2018, https://cns-snc.ca/media/history/fifty_years/foster.html

Bibliography

"Activities of Japan Atomic Energy Commission in 1956." *Atoms in Japan* 1, no. 1 (May 1957): 5–7.

Aichi Prefectural Library. "Kaikan kara 50-nen" ("50 Years Since Opening"). Accessed 9 Dec. 2018. https://websv.aichi-pref-library.jp/event/aichitoshokan_chirashi.pdf.

Aldous, Christopher. "Masking or Marking Britain's Decline? The British Council and Cultural Diplomacy in Japan, 1952–1970." In *The History of Anglo-Japanese Relations, 1600–2000: Vol. 5, Social and Cultural Perspectives*. Edited by Gordon Daniels and Chushichi Tsuzuki, 330–351. Basingstoke, UK: Palgrave Macmillan, 2002.

"A-Power Program Not Yet Crystallized." *Atoms in Japan* 1, no. 1 (May 1957): 2–4.

Asahi Shimbun, 13–14 May, 1957.

"Atomic Achievement: Transcript." UK National Archives. http://www.nationalarchives.gov.uk/films/1951to1964/filmpage_atomic.htm.

"Atomic Gamesmanship," Business Notes, *Economist*, 1 Jun. 1957.

"Atomic Power Arrives," *Economist*, 20 Oct. 1956, 257.

"Burasseru Bankoku Haku Nihon sanka keikaku" ("Plans for Japanese Participation in the Brussels Universal and International Exhibition"). *Kōgei nyūsu* (*Industrial Art News*), 26, no. 3 (March-April 1958), 2–31.

Financial Times, 29 May, 1957, 23 Dec. 1959.

Foster, John. "The Development of Nuclear Power." In *Special Symposium: 50 Years of Nuclear Fission in Review*. Edited by Malcolm Harvey. Ottawa, 5 Jun. 1989, Proceedings of the 29th Annual Conference, Canadian Nuclear Society, Vol. 1, Accessed 19 Dec., 2018, https://cns-snc.ca/media/history/fifty_years/foster.html.

Genshiryoku Sangyō Shimbun (*Atomic Industry Newspaper*), 15 April, 1957, 25 Nov. 1958, 25 April 1958.

Genther Yoshida, Phyllis. *U.S.-Japan Nuclear Cooperation: The Significance of July 2018*. Washington, DC: Sasakawa Peace Foundation USA, 2018. Accessed 9 Dec., 2018, https://spfusa.org/research/u-s-japan-nuclear-cooperation-the-significance-of-july-2018/.

Guardian, 2 Dec. 1964.

Haddow, Robert H. *Pavilions of Plenty: Exhibiting American Culture Abroad in the 1950s*. Washington, DC: Smithsonian Institution Press, 1997.

Haddow, Robert Hamilton. "Material Culture and the Cold War: International Trade Fairs and the American Pavilion at the 1958 Brussels World's Fair." PhD diss., 2 vols., University of Minnesota, 1994.

Hidankyō (Japan Confederation of Atomic and Hydrogen Bomb Sufferers Organizations). "Message to the World," 10 Aug., 1956. Accessed 10 Dec., 2018, http://www.ne.jp/asahi/hidankyo/nihon/rn_page/english/message.html.

Hill, C.N. *An Atomic Empire: A Technical History of the Rise and Fall of the British Atomic Energy Programme*. Singapore: World Scientific Publishing Company, 2013.

Hinton, Sir Christopher. "Atomic Power in Britain." *Scientific American* 198, no. 3 (Mar. 1958): 29–35.

Hiroshoma Fukkō Daihakurankaishi Henshū Iinkai. *Hiroshima Fukkō Daihakurankaishi* (*Report of the Grand Exhibition of the Reconstruction of Hiroshima*). Hiroshima: Hiroshima City Hall, 1959.

Hiroshima-shi Sōmukyoku Sōmuka, *Shisei yōran* (Hiroshima City General Affairs Bureau, General Affairs Division). *Shisei yōran* (*Municipal Handbook*). Hiroshima: Hiroshima City Hall, 1959.

Hiroshima Peace Memorial Museum. "Hiroshima no fukkō" ("Reconstruction of Hiroshima"). Accessed 13 Dec., 2018, http://www.pcf.city.hiroshima.jp/Peace/J/pHiroshima4_3.html.

"Japan Backs Harwell." *Economist*, 16 Jun. 1956, 1120.

Japan Times, 19 Oct., 1956, 28 Nov., 1956, 3 Dec., 1956, 14–28 May, 1957, 1–2 Sept., 1958, 22 Nov. 1958.

Jeandron, Michelle. "The Atomium Marks a Half Century," *Physics World* (April 2008), 7.

"Joint Conference Opens," *Atoms in Japan*, Japan Atomic Industrial Forum, 1, no. 1 (May 1957), 23.

Kanari, Ryuichi. "U.S. Used Hiroshima to Bolster Support for Nuclear Power." *Asahi Shimbun*, 25 Jul. 2011.

Kanō, Mikiyo. "'Genshiryoku no heiwa riyō' to kindai kazoku" ("'Atoms for Peace' and the Modern Family"). *Jendaa shigaku* (*The History of Gender*), no. 11 (2015): 5–19.

Krige, John. "The Peaceful Atom as Political Weapon: Euratom and American Foreign Policy in the Late 1950s," *Historical Studies in the Natural Sciences* 38, no. 1 (Winter 2008): 5–44.

Lindee, M. Susan. *Suffering Made Real: American Science and the Survivors at Hiroshima*. Chicago: University of Chicago Press, 1994.

Low, Morris, Shigeru Nakayama and Hitoshi Yoshioka, *Science, Technology and Society in Contemporary Japan*. Cambridge: Cambridge University Press, 1999.

Mori, Kazuhisa. "Atomic Bombing and Nuclear Energy Development in Japan." Japan Atomic Industrial Forum, 1995, typescript, Yukawa Hall Archival Library, Kyoto University. Accessed 11 Dec., 2018, https://www2.yukawa.kyoto-u.ac.jp/~yhal.oj/Mori/Mori1Q/Reports02/IMG_0025.pdf.

Moritaki, Ichirō. *Kaku zettai hitei e no ayumi* (*Steps toward Absolute Negation of the Nuclear*). Hiroshima: Keisuisha, 1994.

Muto, Ichiyo. "The Buildup of a Nuclear Armament Capability and the Postwar Statehood of Japan: Fukushima and the Genealogy of Nuclear Bombs and Power Plants." *Inter-Asia Cultural Studies* 14, no. 2 (2013): 171–212.

New York Times, 5–16 Apr. 1956, 19 May, 1957.

NHK (Japan Broadcasting Corporation). ("Hiroshima: Bakushinchi Genshiryoku Heiwa Riyō Hakurankai") ("Hiroshima: Ground Zero, Atoms for Peace Exhibition"). *ETV tokushū* (*E Television Special Report*), 2014. Accessed 29 Jan. 2020. http://www.nhk.or.jp/etv21c/file/2014/1018.html.

"Nichibei genshiryoku sangyō tenrankai" ("US Japan Atomic Industries Exhibition"). *Genshiryoku Sangyō Shimbun* (*Atomic Industry Newspaper*), 25 Feb. 1957.

"1958-nen Nihon Kokusai Mihon'ichi kara" ("Inspecting Round the Japan International Trade Fair, Osaka"), 1958, technical paper. National Institute of Advanced Industrial Science and Technology, Tohoku Centre. Accessed 29 Dec. 2018. https://unit.aist.go.jp/tohoku/techpaper/pdf/3888.pdf.

Nippon Times, 1 Jan., 1956, 17–19 May, 1956, 2 Jun. 1956.

Nye, David E. *American Technological Sublime*. Cambridge, Mass.: MIT Press, 1994.

"October 17, 1956 – The Day Which Saw the Beginning of a New Age," *The Illustrated London News*, 27 Oct., 1956, 691–693.

Orr, Joseph. *The Victim as Hero: Ideologies of Peace and National Identity in Postwar Japan*. Honolulu: University of Hawaii Press, 2001.

Petersen, Stephen. "Explosive Propositions: Artists React to the Atomic Age." *Science in Context* 17, no. 4 (Dec. 2004): 579–609.

Puishes, A., D.P. Herron, D.R. Mash and J.W. Webster, *Comparison of Calder Hall and PWR Reactor Types*, prepared for Division of Reactor Development, U.S. Atomic Energy Commission. Redwood City, Calif.: Atomic Energy Division, American Radiator and Standard Sanitary Corporation, 1 Mar. 1957.

Reynolds, Jonathan M. *Maekawa Kunio and the Emergence of Japanese Modernist Architecture*. Berkeley: University of California Press, 2001.

Sapir, Michael and Sam J. Van Hyning. *The Outlook for Nuclear Power in Japan*, Reports on the Productive Uses of Nuclear Energy. Washington DC: National Planning Association, June 1956.

Sasaki-Uemura, Wesley. *Organizing the Spontaneous: Citizen Protest in Postwar Japan*. Honolulu: University of Hawaii Press, 2001.

Schenck, Hubert G. "Impact of Science in East Asia." *Bulletin of the Atomic Scientists* 14, no. 7 (Sept. 1958): 273–275.

Tanaka, Yuki. "'The Peaceful Use of Nuclear Energy' and Hiroshima'" in Yuki Tanaka and Peter Kuznick, "Japan, the Atomic Bomb, and the 'Peaceful Uses of Nuclear Power'." *Asia-Pacific Journal: Japan Focus* 9, 18, no. 1 (2 May, 2011). Accessed 13 Dec. 2018, https://apjjf.org/2011/9/18/Yuki-Tanaka/3521/article.html.

Times, 17 Nov. 1956, 1 Apr. 1957, 14–29 May, 1957, 31 Aug. 1958, 27 Nov. 1958, 23 Dec. 1959.

Totten, George O. and Tamio Kawakami. "Gensuikyō and the Peace Movement in Japan." *Asian Survey* 4, no. 5 (May 1964): 833–841.

Tsuchiya, Yuka. "The Atoms for Peace USIS Films: Spreading the Gospel of the 'Blessing' of Atomic Energy in the Early Cold War Era." *International Journal of Korean History* 19, no. 2 (Aug. 2014), 107–135.

Twigge, Stephen. "The Atomic Marshall Plan: Atoms for Peace, British Diplomacy and Civil Nuclear Power." *Cold War History* 16, no. 2 (2016): 1–18.

US CIA. "Participation of the Sino-Soviet Bloc in International Trade Fairs and Exhibits." Intelligence Memorandum, IM-430, Office of Research and Reports, 1 May, 1956.

US National Archives and Records Administration, College Park, MD. RG 59, General Records of the Department of State; RG306, Records of the USIA; RG 469, Records of U.S. Foreign Assistance Agencies.

Washington Post, 15 Feb. 1955.

Watanabe, Seiki. "The Road to Atomic Power." *Japan Quarterly* 5, no. 4 (1 Oct., 1958): 418–425.

Williams, Roger. "British Nuclear Power Policies." In *Energy Economics in Britain*. Edited by Paul Tempest, 35–58. London: Graham and Trotman, 1983.

Yomiuri Shimbun, 1 Jan. 1956, 14 May, 1956, 18 Apr. 1958, 11 May, 1958.

Zwigenberg, Ran. *Hiroshima: The Origins of Global Memory Culture*. Cambridge: Cambridge University Press, 2014.

CHAPTER 8

Seeing Reactors at Tōkai-mura, Trade Fairs, Department Stores and in Films: 1957–1971

Introduction

With the commencement of Japan's atomic energy program in 1956, more Japanese people were able to experience seeing a nuclear reactor, not only in model form at exhibitions, but in films and by visiting reactors that were under construction and built at Tōkai-mura and elsewhere. It was found that a visit to a reactor could lead to positive changes in attitudes towards nuclear power. By providing information on how nuclear reactors worked, the reactors and their visitor centres helped the Japanese overcome their fears of nuclear radiation and understand how the reactors were helping to meet Japan's future energy needs. A trip to a reactor, often located at a considerable distance from Japan's major cities, helped to make "nuclear power" more visible to the school students, specialists and members of the general public who made the effort to visit and see for themselves.[1] For young Japanese, the trips became a memorable corporeal experience which was formative in their sense of national identity.

Opportunities to visit nuclear facilities emerged against the backdrop of drawn-out negotiations for Japan's purchase of a British Calder Hall type power reactor. This served to increase Japanese interest and also encouraged the USA to showcase what they could offer in terms of reactors. The Third Tokyo International Trade Fair (4–22 May 1959) provided the US Atomic Energy Commission with an opportunity to show the Japanese a working training reactor and a full-scale model of part of the American

M. Low, *Visualizing Nuclear Power in Japan*, Palgrave Studies in the History of Science and Technology,
https://doi.org/10.1007/978-3-030-47198-9_8

Shippingport power reactor. With the formal decision to place an order with the British General Electric Co. for an improved Calder Hall type reactor in late 1959, the debate abated but the promotion of nuclear power continued. Although it had been hoped that the Calder Hall type reactor would be completed in early 1964, it would not come on line and achieve full power production until 1966.

Before that happened, the Japan Atomic Power Company (JAPC) announced in May 1963 that Japan's second nuclear power plant would be an American light-water reactor, a General Electric designed boiling water reactor (BWR) with an electric output of 320 mW. Construction began at Tsuruga, Fukui prefecture in 1966. The Japan Atomic Energy Research Institute (JAERI) completed its own light water demonstration reactor, also a BWR. It was referred to as the Japan Power Demonstration Reactor (JPDR) and successfully generated electricity in October 1963, three years before the Calder Hall reactor.

Meanwhile, JAERI and its scientists were striving to use indigenous technology but Japanese companies sought to fast-track nuclear development and chose to import technology instead.[2] In Japan like elsewhere, there was great interest in light-water reactor technology with interest spread between BWRs and PWRs (pressurized water reactors). GE and Westinghouse dominated the commercial nuclear power industry with their light-water reactors.[3] The first reactor (Unit 1) to be built at the Fukushima Daiichi Nuclear Power Plant was, like in Tsuruga, a GE designed BWR but this time with a 30 per cent larger output.[4] All this development occurred against the background of ongoing promotion of civilian nuclear power in Japan including at the actual reactor sites.

As we shall see in this chapter, the Japanese could go beyond seeing reactors and models of reactors in exhibitions such as the Tokyo International Trade Fair and in department stores and visit the nuclear village itself, Tōkai-mura which became a sightseeing destination. In addition, promotional films were commissioned to record the construction of reactors and these serve as useful resources for understanding how the Japanese public were encouraged to see nuclear power in an optimistic light. In this chapter, we will discuss not only the Trade Fair and department store exhibitions but also the films taken at Tōkai-mura and in Ōkuma and Futaba, Fukushima prefecture.

Visiting Tōkai-mura

On 27 May 1957, just two weeks after the US-Japan Joint Atomic Industrial Forums Conference opened in Tokyo and the ensuing commotion about the merits of the Calder Hall reactor were aired in the international press, the Japanese novelist Takeda Taijun visited Tōkai-mura along with a staff member of the literary journal *Chūō kōron* (*Central Review*) as part of a reporting assignment. As he wrote in his article for that periodical published in July 1957, he accepted the task because there was something fascinating to him about things "atomic" whether it was atomic physics or the generation of atomic energy. It was akin to how natives might see the use of magic with a sense of respect, admiration, and surprise. He had read a newly published translation into Japanese of Kenneth Jay's book *Calder Hall: The Story of Britain's First Atomic Power Station* (1956)[5] and admired how Jay calmly wrote about how the project overcame a series of challenges.[6]

They caught a direct train from Tokyo to Tsuchiura and arrived within two hours, after which they rented a hire car and reached their destination in forty minutes. Takeda noted the beautiful forest of pine trees in the area and quipped that if it hadn't been for them, the facilities would have appeared rather vulgar like Las Vegas. That evening, they were briefed by JAERI director Sagane Ryōkichi at their hotel. Sagane explained how it was a core group of some six or seven young scientists between twenty-three and twenty-five years of age who were instrumental in establishing JAERI. Takeda observed a couple of these young scientists working with a visiting American engineer, Cochrane who was about thirty years old, communicating in the water-boiler type reactor control room using simple Japanese expressions. The two Japanese would later train their colleagues. Cochrane was most likely sent by the Atomics International Division of North American Aviation that sold the 50 kW reactor to the Japanese.[7]

What got in the way of their work was what Takeda called "sightseers" who visited JAERI. They could be divided into three types: A, B and C. Members of group A were made up of those who needed to speak directly with the researchers on detailed matters such as university specialists and people dealing with budgetary matters. Group C included young boys and others who understood little of what they saw but were visiting as part of an excursion that might include a picnic. They could be left to sightsee the facilities and just absorb a sense of scientific spirit. But what was more problematic was group B who were important to the research institute and came armed with fancy letters of introduction. They might include company presidents and office bearers who while not having anything to do with scientific research, still required some scientific explanation. Takeda pitied the young researchers who were taken away from their work and had to deal with such visitors, like a bus tour guide, providing an introduction to the facilities. Takeda thought that in future, they could be replaced by a tape-recorded guide to JAERI or perhaps a retired employee could be employed in such a role. He also mused with some foresight that in future, a director might not need to be a natural scientist but rather could be a government policymaker or corporate manager![8]

The following day, they visited the JAERI facilities at Tōkai-mura, what Takeda referred to as "Genshi-mura" ("Atomic Village"). They started the tour with the water-boiler reactor building, following the so-called "visitor course" through it. They peered through a long window in a concrete wall which enabled them to see inside. Across from the window was another which showed the control room where two Japanese researchers were speaking with Cochrane. The control room was located directly above the reactor. They then returned to the visitor course and inspected the research laboratory followed by the building housing the 6 MeV Van de Graaff accelerator. Takeda could hear bombing from a nearby American base which sent vibrations through the laboratory. That was more of a concern to him than any potential earthquake![9]

The director took Takeda and his colleague to his office where newspaper journalists and cameramen were gathered to see him formally issue a receipt for uranium fuel that the government transferred to JAERI. The uranium had been under police escort on its way there. The media then proceeded to the water-boiler reactor building where the fuel was stored in an underground safe. From that point on, responsibility shifted from the government to JAERI. Takeda's visit served to acquaint him and his readers not only with the facilities but also with many of the issues relating to its establishment at Tōkai-mura, the training of technical personnel, and the need to engage the media to promote JAERI to the public.[10]

The water-boiler reactor came to be known as the Japan Research Reactor 1 (JRR-1) and would reach criticality later that year. It was followed by JRR-2 which was a CP-5 type American reactor with a thermal output of 10,000 kW. Almost two years after Takeda visited, a group of Japanese engineers, researchers and students who had attended the spring meeting of the Iron and Steel Institute of Japan, also visited JAERI. In his account of the 2.5 hour visit on 4 April 1959,[11] Yamada Shigeru from Sumitomo Metal Industries confessed that he didn't understand much of the detail of explanation that he received about the JRR-1 in its control room, but noted that the exterior of the JRR-2 was almost complete and the interior was in the final stages. Meanwhile, preparations were underway for the JRR-3 swimming pool reactor, the so-called first national, domestically made research reactor. Part of the pine forest had been cleared and steel frame erected. He was somewhat disappointed by the tour but was impressed by JAERI efforts to prevent the spread of radioactive dust and pollutants, as well as air purification at the facilities. It was beyond what they had imagined. On their way back to Mito station, the sightseeing bus went along the coastline and the smell of the seashore relieved their tiredness from the tour.[12] The JRR-2 reactor that Yamada had observed would attain criticality in October 1960 and the 10 mW JRR-3 reached criticality in 1962. The latter was a milestone of sorts as the idea of constructing their own research reactor had been a long-term goal of Japanese scientists and technicians.[13]

About a year after Yamada had visited, a *Yomiuri Shimbun* reader wrote to the newspaper's advice column seeking information about visiting atomic energy facilities at Tōkai-mura. The newspaper reported that JAERI, as Japan's only atomic energy centre, had attracted international attention. There was an average of 1000 sightseers per day. On busy days, they could reach 3000. Visitors could freely stroll around the grounds and view the JRR-1 water-boiler reactor and the Van de Graaff accelerator. The CP-5 JRR-2 reactor was still under construction at the time of publication in April 1958. The most popular facility among visitors was JRR-1. In order to service the many sightseers, a visitors room had been constructed which enabled them to peer in the reactor room through a glass window. On pressing a button, they could hear a tape-recorded explanation that a first year senior high school student could understand. Also, to give visitors a sense of the whole, a model of the reactor was also on view. Scientists, engineers and researchers could apply to visit Saturday afternoons and enter the reactor room where they would receive more detailed explanations from JAERI personnel.[14]

The newspaper also reported that Ibaraki prefecture was planning to start a sightseeing bus service to take visitors to JAERI and new tourist sights. This would include the Museum of Atomic Energy in Mito where some exhibits from the *Yomiuri*-sponsored *Atoms for Peace* exhibition were on display, along with new reactor models.[15] In this way, the *Atoms for Peace* exhibition left a lingering legacy with items spread strategically between the Hiroshima Peace Memorial Museum and Mito, not far from Tōkai-mura.

School Children at Tōkai-mura

From 21–22 November 1958, a ten-member inspection team from the Tokyo Metropolitan Education Agency's section responsible for school excursions appraised Ibaraki prefecture's "Science Tourism Course". A round-table discussion with some members of the team and around twenty people involved in the prefecture's tourism industry was held in the first evening. The audience included representatives from the prefectural government, the local tourist association, Mito rail management agency, and the Japan Travel Bureau. An Ibaraki prefecture official named Takahashi Seisaburō who had shown the inspection team around JAERI at Tōkai-mura and the Museum of Atomic Energy (Genshiryoku-kan) at Mito earlier that day, explained how JAERI and the centre were key features of the Science Tourism Course. He sought the delegation's advice as to whether the course would be suitable for school excursions and if so, what needed to be done in terms of improvements in order to be approved. Much was at stake as the Tokyo Metropolitan Education Agency was responsible for 2 million students. The meeting noted that the Science Tourism Course had not been included in the guidebook for school excursions for Tokyo schools.[16]

Sano Masao, who was one of the people responsible for writing the guidebook and a member of the inspection team, pointed to the lack of promotional activities aimed at Tokyo schools by Ibaraki prefecture and advised that more needed to be done to promote travel to the north of Tokyo, especially in light of what the inspection team had seen. He noted how this time, the Ibaraki Tourism Association had consulted the distinguished educationalist Yokose Gorō and made more efforts to promote the course to schools. Sano also suggested that they lobby the Japan Travel Bureau and Japan National Railway (JNR) to extend the Natural Science Train which took school children to places of educational interest, to Tōkai-mura as well. It already went to Tsukuba.[17]

He and the inspection team were grateful for all the photographs and guidebooks that they had received, however he suggested that rather than overwhelming visitors with so much material that they couldn't read, it would be better for them to invest some funds in the production of an authoritative guidebook with a small number of good, high-quality photographs. In addition, it was important for there to be good tour guides for teachers and students.[18]

The educationalist and team member Madarame Fumio noted how 4 million yen had been spent building the Ibaraki Prefectural Museum of Atomic Energy at Mito that had opened in April 1958. He praised the displays and jokingly added that after having gone around the museum once, they knew the equivalent of what the Nobel Prize winner Yukawa Hideki would probably have known when he entered primary school! Thanks to the museum, visitors learnt what atomic energy was. He thought that, given that Japan was standing at the door to the age of atomic energy, it was a good opportunity to promote the Science Tourism Course. Access to the area had improved with better roads and by public transport, it was two hours away from Tokyo. In his opinion, a combined effort to promote the area was needed. This should include the production of materials such as brochures for children and educational slides.[19]

Sano noted that in recent times, "technical tourism" had grown in popularity. Rather than just going to the remains of famous historic sites or scenic locations, people, especially in Tokyo, were also keen to see modern industry at work and watching new industries emerge. Such locations were lacking in Nara, Kyoto, Kamakura and even in Chiba. He felt that JAERI would add a new element to school excursions that the Tokyo metropolitan area lacked. Team member Suzuki Hideo suggested that the Science Tourism Course had two features that were unique to it: pure science facilities (JAERI and the Museum of Atomic Energy) and the fact that even primary school students could be taken on the trip as there wasn't anything dangerous. Thus, by 1958, there was a perception that atomic energy was safe and that reactors were sites that could be both educational and of interest to tourists.[20]

Suzuki felt that the route might fit the needs not only of science education but also of moral education, both of which the Ministry of Education was keen on promoting. A route with a focus on Mito would meet the requirements of moral education and one centred on Tōkai-mura met the needs of science education. In that regard, Sano added that the Jōyō Meiji Memorial Hall that commemorated the Meiji emperor and was on the route would be good for primary and junior high school students.[21]

Fellow team member Yokota Hiroyuki agreed with Suzuki that the course's strength was that it combined both elements. He did, however, have one criticism regarding the Museum of Atomic Energy at Mito. It was not clear what audience it was aimed at. He felt that its present content was beyond the comprehension of primary school students and that more thought should be given to content that they could more easily understand and study.[22]

Inumaru Katsuyoshi from the Japan School Trip Association (Nihon Shūgaku Ryokō Kyōkai) stated at the meeting that the year saw a boom in interest in school excursions, with the Ministry of Education planning to allocate 310 million yen in support and trains being built specially for school trips. In Tokyo alone, there were approximately 2 million school students. Such students would travel to Mito in spring and autumn, and on their return, would tell their parents and neighbours about Mito, thereby spreading the word of what was there. When they became adults themselves, they would take their families and visit again. So in a way, the school students would become pioneers of sorts in promoting tourism. He did make a cautionary comment that in Tokyo, there had been those who visited JAERI and complained that they couldn't understand anything and that going to such a place was meaningless. However unlike them, the inspection team had first visited the Museum of Atomic Energy in Mito, received a kind explanation from the museum director, and then proceeded to JAERI from there. As a result, they were able to deepen their understanding of JAERI much more than was expected.[23]

Yokota Hiroyuki wrote a report on what they had experienced which was published alongside a record of the discussion. He felt that in terms of substantive content within the framework of the promotion of science and technology, it was the ability to view atomic energy related sites that made the difference. On the other hand, in terms of the spiritual and national frame, and interest in promoting moral education, one can see the ardent love of the motherland of the Mito clan. At a time of democracy, more than ever there are things that can be learnt from Mito clan traditions. These contrasting features have come together by accident but he felt that the route would be valued for these two things, in addition to other attractions such as the natural surroundings.[24]

He thought that the Museum of Atomic Energy and JAERI were wonderful. At the museum in Mito city, they saw films and slides, as well as models of the reactor which provided highly useful orientation for when they visited Tōkai-mura. At the actual site, they had to look at things through glass and from a distance, so without such preparatory knowledge, the interest of visitors would be halved. At the museum, it was important to prepare easy-to-understand material for a range of visitors from children through to the elderly. He felt that it was important to convey to boys that the dream of a new society built around atomic energy-related science and technology was within their grasp. Even for adults like himself, too, being given big dreams about the future helps make life more enjoyable and meaningful.[25]

For Yokota, the trip to the Museum enabled students to retain some factual information that would be beneficial when they visited the nuclear facilities. The Museum could convey content to students and also affect their values. Students would develop empathy regarding efforts to introduce nuclear power in Japan and view the future in a more positive light.

To enhance the overall visitor experience of Tōkai-mura, Yokota recommended that the prefecture and its various organizations and bodies promote what they had to offer in a more coordinated and rational way. They needed to find the most effective way to do this and to determine a walking course that could be done most cheaply, quickly and conveniently. The relationship between the science tourism route and the Tsukuba route would have to be thought through more carefully.[26]

Within six months of the inspection team's visit, it was announced that JNR East's Natural Science Train would include a new spring school excursion route to Mito and JAERI at Tōkai-mura that ran week-days.[27] In addition, Ibaraki prefecture produced a comprehensive guidebook expressly aimed at the school excursion market that was available from the Ibaraki Prefectural Tourism Association in Mito and its office in the Kokusai Kankō Kaikan building located at Marunouchi, Tokyo. The publication included words of greetings from various officials. A later edition included a message from Sano Masao who once more linked what he saw on the Ibaraki tourism route to science and moral education. He wrote that the various history-related sites at Mito would enrich the heart and spirit of Tokyo children. Also the new sightseeing destination of JAERI would give hope and dreams to people who are moving about in confusion, going every which way in their lives in the big city.[28] Thus, the dream of nuclear power was seen as giving the Japanese people a sense of direction and the location near Mito would ensure that young Japanese visitors would enjoy a sense of both tradition and modernity when visiting the facilities. The description of Tōkai-mura in the guidebook also emphasized how various artefacts had been excavated from historic ruins in the village and how it had been reborn as the latest "*genshi no mura*", an atomic village.[29] In this way, students would experience the sites symbolically as part of a larger spatial and temporal story.

The guidebook provides a brief history of JAERI and refers to how in April 1966, a Nuclear Power Dissemination Centre had been established at Tōkai-mura as a type of orientation centre for visitors. The Museum of Atomic Energy had shifted there from Mito, and within the grounds a sightseeing centre and film room had also been built.[30] As for the JAERI reactors, the guidebook outlines a series of milestones. JRR-1 had reached criticality on 27 August 1957, with JRR-2 doing so on 1 October 1960 and JRR-3, the first Japanese-designed, national reactor on 12 September 1962. This was followed by the JRR-4 swimming pool reactor which reached criticality in October 1964. Construction of a Japan Power Demonstration Reactor (JPDR) was completed in December 1960 and went critical in August 1963, before the Calder Hall reactor. The JPDR was successful in becoming the nation's first reactor to generate nuclear power. The guidebook also explained that power generation by Japan Atomic Power Company's improved Calder Hall reactor had been delayed until January 1966.[31]

The promotion of school trips to Tōkai-mura in the late 1950s and 1960s helped to establish that destination as a centre of Japanese science and technology and contributed to national narratives about Japan's future that would be on display at major events such as the 1964 Tokyo Olympics and the 1970 Osaka Expo. This will be explored in greater detail in the next chapter. Meanwhile, there would continue to be exhibitions promoting atoms for peace in consumer-friendly spaces such as department stores.

Reactors at Department Stores

The connection between nuclear power and everyday life was reinforced by sponsorship by the Japan Atomic Industrial Forum (JAIF) and its associated Japan Foundation for the Peaceful Uses of Atomic Energy in Tokyo department stores. On the occasion of the JAIF's third anniversary, an exhibition was held in Tokyo's Tōyoko Department Store in Shibuya, 17–22 March 1959. It went on to tour other Japanese cities.[32]

At the main entrance to the exhibition was a spectacular panorama of a nuclear-powered city of the future. Visitors then saw a colour slide show that introduced them to nuclear power, followed by photographs and text panels describing the history of its development. The story included mention of historical figures such as Wilhelm Röntgen, Marie Curie, Albert Einstein and Japan's own Nagaoka Hantarō who proposed a Saturnian model of the atom. In a section of the exhibition on reactors and power, visitors were introduced to a range of research reactors just completed or soon to be built at JAERI at Tōkai-mura: a water-boiler type reactor, the CP-5 type research reactor, and swimming pool research reactors. Also on display was a model of a nuclear-powered ship and a 1:100 scale model of the improved Calder Hall-type power plant that would be built at Tōkai-mura by the JAPC.[33]

The following year, an *Atoms for Peace* exhibition was held for the fourth time in Tokyo at the Seibu Ikebukuro Department Store.[34] In this way, the Japanese consumer was introduced to the various reactors being constructed in Japan at Tōkai-mura and the message was continually reinforced of how nuclear power was the way of the future and would underpin Japan's prosperity. Such exhibitions created visual experiences for shoppers who increasingly made the connection between the lifestyle and appliances that they aspired to own and the electric power that was needed to run them.

Third Tokyo International Trade Fair, 1959

For those not able to make the trip to Tōkai-mura to view the progress being made at the Japan Atomic Energy Research Institute, yet another option was to see what the USA had to offer at the Third Tokyo International Trade Fair, 5–22 May 1959. The eighteen-day fair was open to the public for eight days during which time it was hoped that the fair would attract some 1.5 million visitors.[35] Total admissions during the fair exceeded this with more than 1.87 million visitors admitted, including 5700 foreign buyers.[36]

The USA boasted the largest government exhibit with the Department of Commerce and the US Atomic Energy Commission each having their own pavilions in a prime position on the right at the entrance to the trade fair venue at Harumi. These twin pavilions were located together but separated by a courtyard. Each pavilion consisted of an airy space frame structure made of aluminium that was suspended from a 118-foot mast by use of suspension cables. The traffic plan was that visitors would first view the Department of Commerce pavilion and then progress through the courtyard to the AEC pavilion.[37] Sheldon Wesson, writing for *The Japan Times*, praised the atomic energy pavilion. Technical information was displayed in both English and Japanese. And in terms of display:

> The bubbling tubes and scale models are sure-fire dramatizations. These is even a splashy sort of decoration done in colors with lights and glass and plastic, represented as the "spirit of atomic energy," or whatever.[38]

Japanese visitors had the opportunity to view a working, training reactor designed by Advanced Technology Laboratories (ATL), a division of American Radiator and Standard Sanitary Corporation which was based at Mountain View, California. Billed as the second atomic reactor to become critical in Japan and the very first in Tokyo itself, the reactor was pre-set to produce a negligible amount of power: one-tenth watt. Nevertheless, it afforded many members of the general public in Japan the opportunity to observe a reactor in operation.

The UTR-1 research reactor was graphite and water moderated. The reactor had been approved by the Japan Atomic Energy Commission as being safe and there had been hopes that the company's agents, American Commercial Inc. of Tokyo, that the US$125,000 reactor could be sold in Japan for research and training purposes, not least because the shield tank of the reactor had been constructed by a local firm, the Chiyoda Chemical and Engineering Co. of Tokyo. Harold J. Miller from ATL had supervised installation in Tokyo and thought that the safe, low-cost reactor could be used by universities to teach students studying nuclear physics and reactor engineering, as well as cognate areas such as chemistry, agriculture and medicine.[39] However, after being shown at the *Atoms for Peace* exhibit at the Tokyo International Trade Fair, it appears that the reactor was later shipped to Cairo, Egypt and Lahore, Pakistan for exhibition in 1960.[40]

The Financial Times noted that some of the items on display had been shown at the United Nations International Conference on the Peaceful Uses of Atomic Energy held in Geneva in September 1958.[41] Visitors to the US AEC pavilion could view a full-scale model of part of the American Shippingport power reactor that had been on display at Geneva.[42] It was hard to miss, located at the entrance to the building. Another attraction was a model of the world's first nuclear powered passenger-cargo ship, the N.S. Savannah. The US$31 million ship was powered by a pressurized water reactor and could cruise for three years without the need to refuel.[43]

Some Japanese visitors would have recognized the "magic hands" exhibit from previous exhibitions held in Japan. But this time, there was a difference. The display showed the public how nuclear scientists could safely handle highly radioactive materials safely from a shielded room via remotely controlled "hands" or manipulators but in contrast to the past where the hands relied on mechanical linkage to connect the worker to the materials, now they utilized electronic control to transmit signals for operation.[44]

A display of radioisotopes in the AEC pavilion highlighted how Japan was a major consumer of American radioisotopes. Visitors could see them being used in a working model of a scale that used radioisotopes to weigh freight cars, as well as in light sources that could emit light for as long as ten years without batteries or electric light bulbs. There were other displays of industrial uses and films explained how radioisotopes were being used particularly in medicine, as well as in biology and agriculture.[45] But despite all these displays, what was more exciting to the Japanese was the fact that power reactors would be built in Japan.

Science Films: Reactors under Construction

The film *Tōkai hatsudensho no kensetsu: dai ichibu* (*The Construction of the Tōkai Power Plant: Part One*) was released in 1962 and shown to the National Diet, government agencies, companies and to local citizens. Visitors to the construction site were also shown the film.[46] With completion of construction in 1965, a follow-up forty-five-minute colour film entitled *Tōkai hatsudensho no kensetsu kiroku: dai nibu* (*A Record of the Construction of the Tōkai Power Plant: Part Two*) (1966) was sponsored by JAPC and produced by the Iwanami Eiga Seisakusho (Iwanami Productions). Iwanami mainly produced educational and promotional films.[47] A forty-minute film, *Genshiryoku hatsuden no yoake* (*Dawn of Nuclear Power Generation*) (1966) was also organized by the No. 1 Atomic Energy Industry Group and produced by Tokyo Cinema. It won numerous prizes and accolades. Sponsored by the Dai-Ichi Ginkō bank, the film involved the cooperation of numerous organizations both public and private including JAPC, the Tokyo University Institute for Nuclear Study and the Rikkyō University Atomic Energy Research Institute.[48]

Dawn of Nuclear Power Generation begins with a somewhat surreal scene showing the beach at night at Tōkai-mura. In the background, strangely lit structures are being erected. It is November 1960 and we are told that construction of Japan's first nuclear power plant is underway. Building the power station has required the development of new technologies and a high level of precision. It promises to be the first step in opening up a new way of life in a country that is lacking in resources. We see a crate arriving from Great Britain with equipment. Before long it is

May 1963 and the reactor vessel is being worked on and by 1964, the graphite moderator sleeves for the canned fuel rods are being fitted. The graphite serves to slow down the fast neutrons that are emitted. We are told that the hexagonal sleeves were especially designed for an earthquake-prone country like Japan. The film shows the fuel rods being transported to Tōkai-mura. We see a cutaway image of a fuel element clad in Magnox, an alloy consisting mainly of magnesium and small amounts of aluminium and other metals. Each of these clad elements was 80 centimetres long and 20 kg in weight. In April 1965, they are inserted into the reactor. We are shown a pool that is used to cool and safely store spent fuel.

Close to the end of the film, we see that construction of the nuclear power station has finished. A couple of foreign experts meet with the Japanese and the film emphasizes how nuclear power generation is a completely new venture for Japan. Thanks to the efforts and strong commitment of Japanese engineers, the power station was making history. On 4 May 1965, the sun rises and after five years, Japan's first nuclear power reactor reaches criticality. Everyone in the control room celebrates and the event makes newspaper headlines. We are told that, through reactors, the Japanese have effectively harnessed atoms for peace and opened up a new future. The film ends with aerial footage of the Tōkai nuclear power station which has pioneered the way forward to a new era. Thanks to Japan's nuclear technology, Japan's atomic age had begun.

The film serves as a chronicle of various milestones during construction but makes no mention of the problems that were experienced on the way to completion of the power plant. Almost all of the faces of the workers are Japanese and Japan seems to be appropriating the British nuclear technology and portraying it as their own. What's more, the film seems to naturalize nuclear power and down play any danger by likening it to human exposure to cosmic rays which are discussed halfway through the film. Viewers are provided with much technical detail. This tends to give the impression that construction is purposeful and precise.

Meanwhile, the Tokyo Electric Power Company (TEPCO) had been conducting surveys and investigations into the feasibility of building a nuclear power plant at a location in Fukushima. TEPCO commissioned Nichiei Kagaku Eiga Seisakusho (Nichiei Science Film Production Company) to produce a documentary entitled *Reimei: Fukushima Genshiryoku Hatsudensho kensetsu kiroku, chōsa-hen* (*Dawn of a New Age: A Record of the Construction of the Fukushima Nuclear Power Plant, Preliminary Surveys*) (1967). This twenty-six-minute, colour film made the case for the construction of the power plant, recording the milestones in its development. The reactors built at what became known as the Fukushima Daiichi Nuclear Power Plant are all boiling water reactors (BWRs), a type of light-water reactor. These require water as both the moderator and primary coolant so proximity to water was crucial in site selection. The film depicts the preparation for the construction of the first of six GE designed boiling water reactors (BWRs) to be built at the site.

The film begins rather ominously with a shot of frothing, violent waves on the seashore then the camera moves to the expansive, high cliff face along the shore. Perhaps the message is that unruly nature would be tamed. The towns of Ōkuma and Futaba in Fukushima prefecture were chosen as the proposed site. About five minutes into the film, we see a building sign and are told that the Tokyo Electric Power Company (TEPCO) Fukushima Research Institute was established in 1964. Initial surveys and examination of the site are carried out and over half way into the film, we see the institute transform into the Fukushima Nuclear Power Construction Preparation Office about a year later in December 1965.

We see clearing of the land and the building of a sea wall. About nineteen minutes into the film we view what would later turn out to be a crucial stage for the preparation of the site. A tractor and tall crane with claw are seen on the bluff and we are told that they would shave 20 metres off the top of the 35-metre high cliff face. The film then reports that in 1967, the preparatory office became the Fukushima Nuclear Power Construction Office. In this way, the film signals yet another milestone having been reached but it would be a fateful one. As it turned out, the bluff was actually reduced by some 25 metres, leaving the site only 10 metres above sea level and vulnerable. That height was convenient for a number of reasons. Setting up seawater pumps to provide coolant water from a height of 10 metres above sea level was cheaper than doing so on higher ground. Lower ground also meant that it was easier to handle items such as the 500-ton reactor pressure vessels that were delivered by boat. What is more, TEPCO dug a further 14 metres to create basement floors for the turbine buildings and emergency diesel generators. Lowering of the bluff enabled the reactors to be built on bedrock.[49] Apparently, GE decided to locate the critical backup systems in the basement floors of the turbine buildings and TEPCO was reluctant to make any changes to GE designs, lacking the expertise and confidence to adapt them to better meet Japanese conditions.[50]

While images of uniformed technical staff predominate, we do see scenes of local residents engaged in farming in what was largely an agricultural area. Towards the end of the film we see locals assisting in construction work, and Japanese technicians meeting foreign engineers, representatives of US General Electric Co. that designed and largely oversaw construction and maintenance of what would become known as the Fukushima Daiichi (No. 1) Nuclear Power Plant. TEPCO had selected GE to supply the complete plant equipment on the basis of a "turn-key" contract. GE in turn subcontracted the Japanese companies Toshiba, Hitachi and Kajima Corporation to build the plant although some equipment was imported.[51] Hitachi had participated in the construction of Japan's first BWR at Tsuruga and there was a subsequent history of the company cooperating with Toshiba and GE to build them.[52] Kajima was involved in building all six BWRs at the Fukushima Daiichi Nuclear Power Plant.[53] For these reasons, the faces of Japanese workers can be seen throughout the film and its sequel and the two films leave the impression that Japanese built the plant despite the fact that the reactor and those that followed at Fukushima Daiichi Nuclear Power Plant were designed by GE.

The first film ends dramatically with waves breaking on the shore and footage showing construction clearly underway. We have some sense of closure in that one stage has ended and that the project is moving on to the next. A thirty-minute, colour sequel film was made four years later to showcase the completed construction of the TEPCO power plant. Entitled: *Reimei: Fukushima Genshiryoku Hatsudensho kensetsu kiroku, dai-ni-bu, kensetsu-hen* (*Dawn of a New Age: A Record of the Construction of the Fukushima Nuclear Power Plant, Part Two, Construction*) (1971), this film, too, opens with footage of strong waves. These act as an element of continuity between the two films, signalling that nature had been harnessed to meet the needs of Japan.[54]

We are told, one minute into the film, that it has been four years since the beginning of construction and that the most important thing is how to safely use nuclear power. In order to do so, great attention has been paid to safety. A scientific explanation of the light-water reactor follows and we are told that nuclear power generation is similar to that of thermal power, only with a reactor replacing the boiler. Viewers are told various dimensions of the equipment to convey the sense of precision and the importance of accuracy. Like a visual diary of construction, the film records various milestones and provides assurances that great care has been taken with the reactor design. Just five minutes into the film, we are told that the reactor building can withstand an earthquake three times that of the Great Kantō Earthquake of 1923.

In the last five minutes of the film, we see the Fukushima Daiichi Nuclear Power Plant Unit 1 reactor reach criticality. Power generation testing is conducted and at this crucial time, for what seems like the first time in the film, we see a foreign face. We are told that the power plant began commercial operation in 1971. Everyone celebrates and we again see a foreign face. In this way, this sequel is a success story, one which seemingly is almost wholly Japanese made whereas the reality was that the reactor was designed by GE and construction overseen by them.

The film notes that a 784 mW Unit 2 reactor was under construction and that Unit 3, it was hoped, would all be Japanese made. As it turns out, Toshiba did indeed supply that reactor but GE would design all six at the Fukushima Daiichi Nuclear Power Plant. We are told, in the film, that TEPCO planned to continue construction until 1985. Despite the promise of ongoing nuclear power development at Fukushima, the film ends on an optimistic note with the message that the new energy arising from the peaceful use of atomic energy would go on to strongly support the lives of everyone. We gain an emotional sense of closure by this feel-good message that speaks to the use of science and technology for the fulfilment of the needs of the Japanese people.

Conclusion

By the 1970s, Japan had entered the atomic age using foreign technology imported from Great Britain and the USA. The Calder Hall reactor at the Tōkai Power Plant had eventually come online in 1966 but due to cost blowouts and design and safety problems it was a disappointment. Japanese utilities turned to American light-water reactors instead.[55] Documentaries nevertheless downplay the foreign origins of technology and highlight what a national achievement their construction represents. They mark the domestication of atomic energy in Japan, underscoring what David E. Nye has described as a "shift away from terror and toward control that is a central characteristic of the technological sublime."[56] Although Nye was writing about the USA, it is also applicable to Japan. A trip to the reactor could be experienced in a number of ways but the messages being evoked seemed to be the same. In the next chapter, we identify some of those discourses at work and how they have contributed to a sense of national identity.

Notes

1. Izumi Wada, "Nuclear Energy Information Centres," *IAEA Bulletin* 32, no. 2 (June 1990): 10–13, esp. 11.
2. Richard J. Samuels, *The Business of the Japanese State: Energy Markets in Comparative and Historical Perspective* (Ithaca: Cornell University Press, 1987), 240.
3. Robin Cowan, "Nuclear Power Reactors: A Study in Technological Lock-in," *Journal of Economic History* 50, no. 3 (Sept. 1990): 541–567.
4. Hada Mikio, Morikawa Tōru, Sazawa Kunito, Nishizawa Hiroshi, Tomita Hitoharu, Takiguchi Yukio, Ike Masae, Ogura Shirō and Itoi Hitoshi, "Tōkyō Denryoku Kabushiki Gaisha Fukushima Genshiryoku Hatsudensho kiki" ("Tokyo Electric Power Co. Fukushima Nuclear Power Plant Equipment," *Toshiba rebyū* (*Toshiba Review*) 24, no. 1 (January 1969): 13–21, esp. 14.
5. Kenneth Jay, *Calder Hall: The Story of Britain's First Atomic Power Station* (New York: Harcourt, Brace and Co., 1956). For the translation, see Kenneth Jay, trans. by Fushimi Kōji, Mori Kazuhisa and Sueda Mamoru, *Genshiryoku hatsudensho: Kōdāhōru monogatari* (*Atomic Power Station: The Story of Calder Hall*) (Tokyo: Iwanami Shoten, 1957).
6. Takeda, Taijun, "Tōkai-mura kenbutsu-ki" ("An Account of a Sightseeing Trip to Tōkai-mura"), *Chūō kōron* (*Central Review*) 72, no. 9 (July 1957): 182–190, esp. 182.
7. Takeda "Tōkai-mura kenbutsu-ki," esp. 182–183.
8. Takeda, "Tōkai-mura kenbutsu-ki," esp. 183–184.
9. Takeda, "Tōkai-mura kenbutsu-ki,", esp. 185–188.
10. Takeda, "Tōkai-mura kenbutsu-ki,", esp. 188.
11. Yamada Shigeru, "Dai 57-kai kōen taikai kōjō kengaku-ki" ("An Account of the 57th Lecture Meeting Factory Tour"), *Tetsu to hagane* (*Iron and Steel*) 45, no. 6 (1 Jun. 1959): 665.
12. Yamada, "Dai 57-kai kōen taikai kōjō kengaku-ki."
13. "Reactor Construction Programme in Japan," *International Atomic Energy Agency Bulletin* 1, no. 1 (Apr. 1959): 5–6; Kiyonobu Yamashita, "History of Nuclear Technology Development in Japan," *Advancing of Nuclear Science and Energy for National Development*, AIP Conference Proceedings, 1659: 020003-1-020003-13.
14. "Genshiryoku Sentā: jiyūni shonai o kengaku dekiru" ("Atomic Energy Centre: Can Freely Sightsee Within the Grounds"), *Yomiuri Shimbun*, 25 Apr. 1958, 3.
15. "Genshiryoku Sentā."

16. "Zadankai: Genken chūshin no Ibaraki no kagaku kankō kōsu o kataru" ("Roundtable Discussion: On Ibaraki's Science Tourism Course Centring on JAERI"). *Shūgaku ryokō*, no. 35 (Feb. 1959): 23–27, esp. 23.
17. "Zadankai," esp. 23.
18. "Zadankai," esp. 23–24.
19. "Zadankai," esp. 24.
20. "Zadankai," esp. 25. An interesting comparison could be made with school tours of the Stanford Linear Accelerator Center (SLAC) in the USA that Sharon Traweek writes about in *Beamtimes and Lifetimes: The World of High Energy Physicists* (Cambridge, Mass.: Harvard University Press, 1988), 23. She found that "most visitors on these tours arrived wanting to be awed rather than informed."
21. "Zadankai," esp. 26.
22. "Zadankai," esp. 27.
23. "Zadankai," esp. 27.
24. Yokota Hiroyuki, "Ibaraki-ken no kagaku kankō rūto o megutte" ("On Ibaraki Prefecture's Science Tourism Route"), *Shūgaku ryokō*, no. 35 (Feb. 1959): 28.
25. Yokota, "Ibaraki-ken no kagaku kankō rūto o megutte."
26. Yokota, "Ibaraki-ken no kagaku kankō rūto o megutte."
27. "Tōkai-mura kōsu kuwaeru" ("Tōkai-mura Course to Be Added"), *Asahi Shimbun*, 17 Apr. 1959, 10.
28. Sano, Masao, "Aisatsu" ("Greetings"), in Ibaraki-ken (ed.), *Ibaraki* (Mito: Ibaraki-ken Kankō Kyōkai, c. 1966), 5.
29. Ibaraki-ken, ed., *Ibaraki* (Mito: Ibaraki-ken Kankō Kyōkai, c. 1966), 52.
30. Ibaraki-ken, ed., *Ibaraki*, 53.
31. Ibaraki-ken, ed., *Ibaraki*, 54–56.
32. "Genshiryoku-ten o Tōkyō de hiraku" ("Atomic Energy Exhibition Will Open in Tokyo"), *Genshiryoku Sangyō Shimbun* (*Atomic Industry Newspaper*), 15 Jan. 1959, 3.
33. "Genshiryoku heiwa riyō tenrankai" ("Atoms for Peace Exhibition"), *Genshiryoku Sangyō Shimbun* (*Atomic Industry Newspaper*), 5 Mar. 1959, 8.
34. "Genshiryoku heiwa riyō tenrankai" ("Atoms for Peace Exhibition"), *Genshiryoku Sangyō Shimbun* (*Atomic Industry Newspaper*), 25 Oct. 1960, 4.
35. Our Own Correspondent, "Tokyo's World Trade Fair," *Financial Times*, 5 May 1959, 5.
36. Tōkyō Kokusai Mihon'ichi Kyōkai (Tokyo International Trade Fair Association) (ed.), *40 nen no ayumi* (*40 Year History*) (Tokyo: The Association, 1996), 29.
37. "U.S. Has Two Pavilions at Current Tokyo Fair," *Japan Times*, 11 May, 1959, 5, 7.

38. Sheldon Wesson, "1959 Exhibit Adopts Educational Approach," *Japan Times*, 11 May, 1959, 6.
39. "Training Reactor on Display," *Japan Times*, 11 May, 1959, 9.
40. *Nuclear Reactors Built, Being Built, or Planned in the United States as of June 30, 1970* (Oak Ridge, Tennessee: US Atomic Energy Commission, Division of Technical Information, 1970), 34, note 40.
41. Our Own Correspondent, "Tokyo World Fair: U.S. Nuclear Exhibits," *Financial Times*, 8 May, 1959, 5.
42. Our Own Correspondent, "Tokyo World Fair: U.S. Nuclear Exhibits;" "Tōkyō Kokusai Mihon'ichi hiraku" ("The Tokyo International Trade Fair Opens"), *Genshiryoku Sangyō Shimbun* (*Atomic Industry Newspaper*), 5 May 1959, 2.
43. "Model of Savannah Featured at Fair," *Japan Times*, 11 May, 1959, 9.
44. "'Magic Hands' Bar Radiation Hazards," *Japan Times*, 11 May, 1959, 9.
45. "Exhibit Shows New Uses for Isotopes," *Japan Times*, 11 May, 1959, 9.
46. Seo, Hanako, "PR eiga ni egakareru genshiryoku: 1950-nendaimatsuha kara 1960-nendai no 'heiwa riyō', 'kagaku gijutsu', 'kindaika'" ("Atomic Energy As Seen in PR Films: 'Atoms for Peace', 'Science and Technology' and 'Modernizatiion' from the Late 1950s Through to the 1960s"), *Shakai jōhōgaku* 4, no. 3 (2016): 47–62, esp. 55.
47. National Science Museum, "*Tokai hatsudensho no kensetsu kiroku: dai nibu*" ("*A Record of the Construction of the Tokai Powerplant: Part Two*"), Sangyō Eiga (Industry Films), accessed 8 Jan. 2019 online at http://sts.kahaku.go.jp/sts/eiga/file/025.htm
48. Science Film Museum, *Genshiryoku hatsuden no yoake* (*Dawn of Nuclear Power Generation*), accessed 8 Jan. 2019, http://www.kagakueizo.org/movie/industrial/87/
49. Reiji Yoshida and Takahiro Fukada, "Fukushima Plant Site Originally was a Hill Safe from Tsunami," *Japan Times*, 13 Jul. 2011, Dow Jones Factiva.
50. Reiji Yoshida, "GE Plan Followed with Inflexibility," *Japan Times*, 14 Jul. 2011, Dow Jones Factiva.
51. Hada Mikio et al., "Tōkyō Denryoku Kabushiki Gaisha Fukushima Genshiryoku Hatsudensho kiki," esp. 13.
52. Masayoshi Matsuura, Kohei Hisamochi, Shinichiro Sato and Kumiaki Moriya, "Lessons Learned from Fukushima Daiichi Nuclear Power Station Accident and Consequent Safety Improvements," *Hitachi Review* 62, no. 1 (2013): 75–80, esp. 75.
53. Kajima Corporation, "History," esp. "Truly Comprehensive Capabilities," accessed 12 Jan. 2019, https://www.kajima.co.jp/english/prof/overview/index.html

54. Science Film Museum, *Reimei: Dainibu, kensetsu-hen* (*Dawn of a New Age: Part Two, Construction*), accessed 9 Jan. 2019, http://www.kagakue-izo.org/movie/industrial/4319/
55. Akira Kinouchi, "30-Year History of Nuclear Power Generation in Japan," *Science and Technology in Japan* 15, no. 58 (1996): 17–21; Samuels, *The Business of the Japanese State*, 240.
56. David E. Nye, *American Technological Sublime* (Cambridge, Mass.: MIT Press, 1994), 234–235.

Bibliography

Asahi Shimbun, 17 Apr. 1959, 10.

Cowan, Robin. "Nuclear Power Reactors: A Study in Technological Lock-in." *Journal of Economic History* 50, no. 3 (Sept. 1990): 541–567.

Financial Times, 5–8 May 1959.

Genshiryoku Sangyō Shimbun (*Atomic Industry Newspaper*), 15 Jan. 1959, 5 Mar. 1959, 5 May 1959, 25 Oct. 1960.

Hada, Mikio, Morikawa Tōru, Sazawa Kunito, Nishizawa Hiroshi, Tomita Hitoharu, Takiguchi Yukio, Ike Masae, Ogura Shirō and Itoi Hitoshi. "Tōkyō Denryoku Kabushiki Gaisha Fukushima Genshiryoku Hatsudensho kiki" ("Tokyo Electric Power Co. Fukushima Nuclear Power Plant Equipment"). *Toshiba rebyū* (*Toshiba Review*) 24, no. 1 (January 1969): 13–21.

Ibaraki-ken, ed. *Ibaraki*. Mito: Ibaraki-ken Kankō Kyōkai, c. 1966.

Japan Times, 11 May, 1959, 13–14 Jul. 2011.

Jay, Kenneth. *Calder Hall: The Story of Britain's First Atomic Power Station*. New York: Harcourt, Brace and Co., 1956.

Jay, Kenneth. *Genshiryoku hatsudensho*: *Kōdāhōru monogatari* (*Atomic Power Station: The Story of Calder Hall*). Translated by Fushimi Kōji, Mori Kazuhisa and Sueda Mamoru. Tokyo: Iwanami Shoten, 1957.

Kajima Corporation, "History," esp. "Truly Comprehensive Capabilities." Accessed 12 Jan. 2019. https://www.kajima.co.jp/english/prof/overview/index.html.

Kinouchi, Akira. "30-Year History of Nuclear Power Generation in Japan." *Science and Technology in Japan* 15, no. 58 (1996): 17–21

Matsuura, Masayoshi, Kohei Hisamochi, Shinichiro Sato and Kumiaki Moriya. "Lessons Learned from Fukushima Daiichi Nuclear Power Station Accident and Consequent Safety Improvements." *Hitachi Review* 62, no. 1 (2013): 75–80.

National Science Museum, "*Tokai hatsudensho no kensetsu kiroku: dai nibu*" ("*A Record of the Construction of the Tokai Powerplant: Part Two*"), Sangyō Eiga (Industry Films). Accessed 8 Jan. 2019. http://sts.kahaku.go.jp/sts/eiga/file/025.htm.

Nuclear Reactors Built, Being Built, or Planned in the United States as of June 30, 1970. Oak Ridge, Tennessee: US Atomic Energy Commission, Division of Technical Information, 1970.

Nye, David E. *American Technological Sublime.* Cambridge, Mass.: MIT Press, 1994.

"Reactor Construction Programme in Japan." *International Atomic Energy Agency Bulletin* 1, no. 1 (April 1959): 5–6.

Samuels, Richard J. *The Business of the Japanese State: Energy Markets in Comparative and Historical Perspective.* Ithaca: Cornell University Press, 1987.

Sano, Masao. "Aisatsu" ("Greetings"). In *Ibaraki.* Edited by Ibaraki-ken. Mito: Ibaraki-ken Kankō Kyōkai, c. 1966.

Science Film Museum, *Genshiryoku hatsuden no yoake* (*Dawn of Nuclear Power Generation*). Accessed 8 Jan. 2019 http://www.kagakueizo.org/movie/industrial/87/.

Science Film Museum, *Reimei: Dainibu, kensetsu-hen* (*Dawn of a New Age: Part Two, Construction*). Accessed 9 Jan. 2019 http://www.kagakueizo.org/movie/industrial/4319/.

Seo, Hanako, "PR eiga ni egakareru genshiryoku: 1950-nendaimatsuha kara 1960-nendai no 'heiwa riyō', 'kagaku gijutsu', 'kindaika'" ("Atomic Energy As Seen in PR Films: 'Atoms for Peace', 'Science and Technology' and 'Modernizatiion' from the Late 1950s Through to the 1960s"). *Shakai jōhōgaku* 4, no. 3 (2016): 47–62.

Takeda, Taijun. "Tōkai-mura kenbutsu-ki" ("An Account of a Sightseeing Trip to Tōkai-mura"). *Chūō kōron* (*Central Review*) 72, no. 9 (July 1957): 182–190.

Tōkyō Kokusai Mihon'ichi Kyōkai (Tokyo International Trade Fair Association), ed. *40 nen no ayumi* (*40 Year History*). Tokyo: The Association, 1996.

Traweek, Sharon. *Beamtimes and Lifetimes: The World of High Energy Physicists.* Cambridge, Mass.: Harvard University Press, 1988.

Wada, Izumi. "Nuclear Energy Information Centres." *IAEA Bulletin* 32, no. 2 (June 1990): 10–13.

Yamada, Shigeru. "Dai 57-kai kōen taikai kōjō kengaku-ki" ("An Account of the 57th Lecture Meeting Factory Tour"). *Tetsu to hagane* (*Iron and Steel*) 45, no. 6 (1 Jun. 1959): 665.

Yamashita, Kiyonobu. "History of Nuclear Technology Development in Japan." *Advancing of Nuclear Science and Energy for National Development.* AIP Conference Proceedings, 1659: 020003-1-020003-13.

Yokota, Hiroyuki. "Ibaraki-ken no kagaku kankō rūto o megutte" ("On Ibaraki Prefecture's Science Tourism Route"). *Shūgaku ryokō*, no. 35 (Feb. 1959): 28.

Yomiuri Shimbun, April 25, 1958.

"Zadankai: Genken chūshin no Ibaraki no kagaku kankō kōsu o kataru" ("Roundtable Discussion: On Ibaraki's Science Tourism Course Centring on JAERI"). *Shūgaku ryokō*, no. 35 (Feb. 1959): 23–27.

CHAPTER 9

Shaping the National Narrative: From Hiroshima to Fukushima and Beyond

Introduction

The Fukushima nuclear disaster that began on 11 March 2011 was a critical moment in the history of Japanese technology. It prompted many to ask why the Japanese people embraced nuclear power and maintained faith in the energy source, despite having witnessed its destructive force in Hiroshima and Nagasaki. This book has sought to answer that question by examining how exhibitions and representations helped achieve a relative consensus regarding the need for the development of nuclear power that ultimately facilitated the transfer of American nuclear technology as well as an improved version of the British Calder Hall reactor.

For the Japanese, we have seen how atomic energy meant more than just Hiroshima and Nagasaki. The dropping of the atomic bomb on those cities did result in widespread opposition to things nuclear but the idea of atoms for peace served to separate the 'good' from the 'bad' at a time when nuclear weapons were nevertheless being secretly deployed in Okinawa, present on US ships and submarines calling at Japanese ports, and being stored in Japan albeit as components.[1]

The previous chapters have focused on the decades through to 1971 which provide a useful window to understanding the promotion of nuclear power as part of Japan's post-war reconstruction and how it came to provide about one third of the nation's electricity supply before Fukushima. Exhibitions discussed in previous chapters have included the America Fair

M. Low, *Visualizing Nuclear Power in Japan*, Palgrave Studies in the History of Science and Technology,
https://doi.org/10.1007/978-3-030-47198-9_9

in Nishinomiya city, 18 March to 31 May 1950; the Kobe Fair, 15 March to 15 June 1950; the *Atoms for Peace* exhibit that toured from November 1955 to August 1957; the *Family of Man* exhibition in 1956 and the Brussels World's Fair in 1958. While the Atoms for Peace discourse was promoted throughout the world, this book has paid particular attention to its Japanese context. We also viewed nuclear films, many in response to the hydrogen bomb testing in the Pacific in the 1950s which provide insights into how US-Japan relations, Japan, its nuclear past and future have been portrayed.

This chapter focuses on truly national events, including the Olympics held in Tokyo in 1964, Expo '70 in Osaka, as well as nuclear accidents culminating in the Fukushima disaster. Together, they helped shape a national narrative about Japan's future and the role of science and technology for both a domestic and international audience. Expo '70 was a grand, futuristic spectacle powered by two new, American-built light-water reactors. It underlined the role of the USA in shaping this discourse and how the Japanese had embraced science and technology after its defeat in World War II. Space technology that was showcased in the US Pavilion at Expo '70 showed there were no limits to human ingenuity and invention. By 1970, we see a shift in discourses on the nuclear, away from what had occurred at Hiroshima and Nagasaki "towards a more optimistic representation of the nation as a champion of peace, science, and technology."[2] Nuclear power seemed to have become less of a technological wonder and the attention of many Japanese turned to moon rock and what could be achieved in space.

The Japanese people were assured that its nuclear power programme was for peaceful purposes but it did provide Japan with the capacity to consider nuclear armament if its leadership deemed it necessary. North Korea looked to the Calder Hall reactor at Tōkai-mura for its own Yongbyon 5MWe reactor which it commenced construction in 1980. Ostensibly to be used for electricity generation, it could easily produce plutonium and its first test of a nuclear weapon in 2006 provided evidence of this. In 2011 the events at Fukushima disrupted the national discourse about the role of nuclear power in Japan. In the final sections of this chapter, we contemplate what the national narrative will be in 2020 at the Tokyo Olympics and the Osaka Expo to be held in 2025. Will history repeat itself?

1964 Tokyo Olympics

In 1959, as the Japanese were deciding to purchase a British Calder Hall type reactor, the International Olympic Committee made the momentous decision to select Tokyo as the site of the 1964 Olympic Summer Games. Chosen on the first ballot, it was a resounding vote of confidence in Japan as well.[3] As the Tokyo governor Azuma Ryōtarō noted, it would be the first time in Olympic history that the games were held in Asia. The games would bring East and West together and contribute to international friendship.[4]

The Japanese director Ichikawa Kon's well-known film *Tokyo Olympiad 1964* (1965) reveals some of the symbolism behind the Tokyo Olympics. The film juxtaposes old and new facets of Japan. It begins with a bright sun and the destruction of old buildings to make way for the new. There is a shot of Greece to show the origins of the Olympics, and then we are transported back to Japan with another shot of a rising sun. The Olympic torch travels through Istanbul, Beirut, Tehran, Lahore, New Delhi, Rangoon, Hong Kong and other cities. We then are reminded of Japan's past with an aerial view of Hiroshima's A-Bomb Dome. Young and old Japanese faces greet blonde foreigners coming off a Pan American jet. The torch is carried through an old Japanese town, tiled roofs of old houses flagging traditional Japan, helped along by a glimpse of *bunraku* puppet heads.

We return to signs of Japan's emerging modernity with the jostling and bustle of Japanese waiting to greet the foreigners in Tokyo for the Olympics. An image of Mount Fuji fills the screen, with what appears to be a vehicle zooming across the foreground. The car signals youth and speed, just like the international athletes visiting what clearly is a modern Tokyo. The raising of flags reinforces the image of an international event, as does the parade of countries in the opening ceremony.

The camera captures the faces of Emperor Hirohito and his son Prince Akihito, as well as the face of an unknown old man overcome with emotion. The opening of the Olympics by the Emperor is followed by the arrival of the final torch bearer, Sakai Yoshinori. Sakai was born on 6 August 1945 in Miyoshi, Hiroshima prefecture, some 60 kilometres from Hiroshima city itself where the atomic bomb had been dropped. Dubbed the "Genbakko" (child of the atomic bomb) for this reason, he represented both the coming of peace to Japan and hope for the future. He referred to himself as a "shūsenkko" ("end of war child").[5] Sakai was a member of the Waseda University athletics club but did not qualify for the Tokyo Olympics. The nineteen-year-old, first-year student in the Faculty of Education at Waseda had been very disappointed at not being selected to run in the men's 400 metres race but would nevertheless get to wear the white running singlet with the red sun and five Olympic rings after all.

Sakai participated in the opening ceremony under a beautiful blue sky at the national stadium in Tokyo on 10 October 1964. Seventy thousand people in the stadium looked on and television cameras broadcast the images to an international audience. Sakai had three minutes in which to receive the torch, run 500 metres, climb some 164 steps to the top and lower the torch into the cauldron.[6] He had a strong stride and flowing form that had prompted his school friends to dub him "kamoshika" ("Japanese antelope") from his junior high school days.[7] He would make the cover of *Sports Illustrated*[8] and be written about in newspapers throughout the world. Although he was an athlete and would go on to participate in the 1966 Asian Games where he won a gold and silver medal in the 4 × 400 m relay and 400 m sprint respectively, his importance lies in his participation in that opening ceremony. It was more symbolic than athletic. He lit the Olympic flame in the main stadium, and pigeons symbolizing peace were released to signal a new Japan. As Ichikawa's film makes plain, Sakai contributed to a narrative about the rebuilding of Japan. When interviewed, Sakai preferred to downplay the significance of the place and time of his birth in his selection as final torchbearer.[9] In a statement that was published in the *Asahi Shimbun* newspaper, he said that it was just a coincidence that he had been born on the same day that tens of thousands of Japanese had become, in one moment, victims of the war.[10] *The New York Times* added drama by describing him as having been born "within sight of the atomic blast" but that was not quite true.[11] Sakai's father was reported as having seen the flash of the atomic bomb from his home village[12] but his son would explain that his father had gone shopping several hours before he was born. He had travelled 70 kilometres to the southern part of Hiroshima prefecture, closer to Hiroshima city where the atomic bomb had been dropped.[13]

As the *Asahi Shimbun* euphemistically put it, it was Japan's first step towards peace.[14] This preference to dwell on the positive rather than negative can be seen in an article published in the evening of the opening ceremony. Whereas some foreign journalists referred to Sakai at the "atomic boy" reminding readers of what had happened at Hiroshima, the *Asahi Shimbun* preferred to shift attention to the future and think of the atomic bomb as the atomic fire that opened the door to a new period of peace.[15] The *Asahi Shimbun* approach reflects the Japanese narrative whereas the American view of the atomic bomb tends to dwell on how it ended the war and saved American lives.[16] While these are different, Igarashi Yoshikuni has suggested that they complement each other as the Japanese stance seems to acknowledge that the atomic bomb was necessary to end the conflict between the USA and Japan.[17]

The selection of Sakai to light the flame that burned throughout the Tokyo Olympic Games was intended, it argued, "to dramatize pacifist Japan's special horror of nuclear weapons."[18] A journalist named Kuwajima, writing for the *Asahi Shimbun*, described Sakai as an earnest, somewhat timid and inconspicuous young man. He gave the impression of being an unsophisticated, senior high school student rather than a university student. Now, his face was appearing on television and in newspapers and he had become a star, despite his prior life having been just the same as any other young Japanese, having been born to a family headed by a salaryman father[19] who, as it happened, worked in the Miyoshi sales office of Chūgoku Electric Power Co.[20] But it was perhaps Sakai's ordinariness that also appealed to the organizers.

In this way, the 1964 Olympics and the media coverage that it generated helped to articulate to an international audience that Japan had rebuilt and recovered. In Ichikawa's film, we see both Sakai and the Hiroshima A-Bomb Dome but there is an imperative made clear in the film and the book, for the Japanese to put all this behind them and to move forward. The Olympics, as Paul Droubie has argued, provided an international stage for Japan to perform its new identity as a peace-loving nation that despite having fallen victim to nuclear weapons was looking to the future and only selectively remembering the past.[21] It was, as Sakai Yoshinori argued, "not just a sports event, but a national program to recover from World War II."[22]

Sakai's lighting of the cauldron was viewed by *The New York Times* as "an expression of the national will for peace divorced from ideology."[23] Yet a month after the opening ceremony, an American nuclear-powered submarine, the Sea Dragon, entered Japanese waters for the first time, calling at Sasebo harbour in Nagasaki prefecture from 12 to 14 November 1964. Its arrival had been postponed so as not to detract from the Tokyo Olympics.[24] It was evidence of intensification of the Vietnam War but also a sign of something more.[25] The relative lack of protests[26] signalled more public acceptance towards nuclear power. In a telegram to the US

Department of State sent on 14 November, US Ambassador to Japan, Edwin O. Reischauer regarded the visit of the Sea Dragon as a "welcome turning point in Japanese public thinking, indicative of considerably more progress toward public acceptance of 'things nuclear' than heretofore had generally been expected."[27] He surmised that this was because the nuclear propulsion of the submarine fell into the category of peaceful use of nuclear power. But he qualified his comments to say that "this of itself cannot be interpreted at this time as reflecting any greater willingness on part [of] Japanese public [to] accept nuclear weapons."[28]

Writing in *The New York Times Magazine* in August 1965, George Packard suggested that "young Japanese are probably less concerned with Hiroshima than are young Americans". He interviewed Takemoto Minoru, a twenty-year-old student at Hiroshima University who was more worried about the bombing in Vietnam than what had occurred in the past at Hiroshima.[29] Packard interpreted the lighting of the cauldron in Tokyo and the circumstances of Sakai's birth in Hiroshima prefecture as showing how Japan had been reborn, phoenix-like from the ashes of Hiroshima. Japan had rejoined the community of nations and "the Olympic flame was, for many, an expression of restored self-confidence and of the pride that had vanished on the U.S.S. Missouri just 20 years ago this week."[30]

In the wake of the success of the 1964 Tokyo Olympics, Japan embraced industrial tourism by establishing a Technical Tourism Plan which sought to showcase both traditional and modern Japan. As part of this plan, the Japan Travel Bureau and financial newspaper *Nihon Keizai Shimbun* created a programme that combined sightseeing tours of Japan's most popular tourist destinations with visits to the facilities of leading companies that could provide a cross-section of Japan's major industries.[31] In this way, the government sought to actively shape the image of Japan as a vibrant, industrial nation. As if to underline this, in 1968, the industrialist and author Takahata Seiichi published his book *Industrial Japan and Industrious Japanese*. In the section on atomic energy, Takahata repeated the argument that the physicist Taketani Mituo had made back in 1952. Takahata wrote how the Japanese were the only people who had suffered the horror of two atomic bombs:

> so they are perfectly qualified to oppose the abuse of horrible atomic power for war, but, at the same time... be second to none in supporting the use of

such a power for peaceful purposes to benefit mankind. Misfortune can be turned into fortune forever.[32]

Expo '70

By 1970, nuclear power generation had made more progress. This was flagged at the futuristic 1970 Expo held in Osaka from 15 March to 13 September. Like the Olympics, it was the first such event to be held in Asia. The Expo's theme was "Progress and Harmony for Mankind." In March 1970, just prior to the opening of Expo '70, *Time* magazine put it on its front cover and declared to the world that "No country has a stronger franchise on the future than Japan."[33] Some Japanese felt that as the first country to experience the full force of the atomic bomb in wartime, it had a right to exploit nuclear fission for peaceful purposes. Some 64 million people would be able to glimpse the prospect of a nuclear powered future over a period of six months, spread out over 330 hectares and at a cost then of $2.9 billion.

The Expo not only celebrated Japan's post-war achievement of rapid economic growth but it also marked the successful commencement of operation of Japan's first American-made light-water reactors (LWRs), both located in Fukui prefecture. Indeed, it flagged many firsts. Japan Atomic Power Company's Tsuruga No. 1 boiling water reactor (a type of LWR), built by General Electric, began operations on 14 March in time for the opening of Expo '70 by Emperor Hirohito that day. The Tsuruga reactor was Japan's first boiling water reactor for commercial use. Not to be beaten, Kansai Electric Power Company's Mihama Nuclear Power Plant Unit No. 1 reactor in Fukui prefecture started trial operations that summer in order to also provide atomic powered lighting for the Expo. Japan's first pressurized water reactor for commercial use reached criticality on 29 July and on 8 August, the anniversary of the dropping of the atomic bomb on Nagasaki, the Expo's electric bulletin board proclaimed that the Mihama Power Plant, some 170 kilometres away, had sent atomic powered electricity to the site, providing "atomic light" to the Expo venue. The reactor had been constructed by the nuclear power division of the American-based company Westinghouse Electric. The press made much of this so-called light of peace which would open up and illuminate the future for the Japanese people, albeit courtesy of American technology.[34]

Samuel Jameson, writing for the *Chicago Tribune* suggested that "most of Expo's art, innovation, experimentation, and fun was being found in the Japanese government and the private industry pavilions." He likened the government's pavilion to five gas tanks but within them he was impressed to find two outstanding tapestries which represented the tragedy and joy of atomic energy, namely the bombing of Hiroshima and the peaceful uses of atomic power.[35] Designed by Kōno Takashi (1906–1999) who was also responsible for the exhibition design within the Japan pavilion, the impressive tapestries measured some 9.2 meters in height and 19.2 meters in width. Entitled "Tower of Joy" and "Tower of Sorrow," they were displayed within what was referred to as the two atomic towers which could be found standing within Hall No. 4, one of Jameson's so-called gas tanks.

Kōno was an important figure in Japan's design scene, having been a member of the design collective Nippon Kōbō from c. 1934 to 1939, working on the renowned propaganda magazine *Nippon*. In addition, he helped design photographic murals displayed in the Japan pavilions at the Exposition Internationale des Arts et Techniques dans la Vie Moderne in Paris in 1937, and the Golden Gate International Exposition (1939–1940) in San Francisco.[36]

The pavilion that Kōno worked on in Osaka consists of five cylindrical structures which represented the petals of a cherry blossom, the official Expo emblem.[37] The aim of the pavilion was to express the heart of Japan and its energy as seen in its past, present and future.[38] Hall No. 1 was devoted to the past, providing visitors with a time tunnel by which to review the history of Japanese culture. Hall No. 2 and No. 3 both dealt with the present. Hall No. 2 focused on "Industries of Japan" and "The Daily Life of the Japanese" and Hall No. 3 told stories on the theme of "The Sun and Water" such as "Japanese Nature and Its Utilization" (including agriculture and forestry) and "Four Seasons of Japan." Halls No. 4 and No. 5 were devoted to the future and it was in Hall No. 4 that Japan's achievements in science and technology were reviewed with the help of a fifteen-panel display showing the progress that had been made. Visitors also entered a dome entitled 'Our World' which had seventeen screens showing photos of daily life of people in seventeen cities throughout the world at that exact time.[39]

Hall No. 4 housed the two atomic towers which were reflected the two faces of atomic energy: the Tower of Atomic Holocaust and the Tower of Atoms for Peace.[40] As the peace activist and atomic bomb survivor Moritaki

Ichirō put it several years later, there seemed to be a consensus that nuclear weapons were bad and associated with death and that Japan could safely and should pursue civilian nuclear power.[41] This way of thinking can be seen in Kōno's tapestry "Tower of Sorrow" which hung in the atomic tower dealing with the atomic bombing of Japan. It depicted a dark mushroom cloud over Hiroshima, the destruction of the city, and the sadness of the victims. It effectively spoke out against nuclear weapons. By 1970, the promise of civilian nuclear power generation had become a reality for the Japanese and the brighter, more positive side of the atom had become apparent. Kōno's tapestry "Tower of Joy" celebrated the awesome power of the atom and how it could improve people's lives through generating electricity and powering nuclear ships, and in saving people's lives through radiation medicine.[42] The large tapestries were woven in the Nishijin district of Kyoto, not far from Osaka. While Moritaki acknowledges that the tapestries reflect public attitudes to nuclear power at the time, he also notes how they were part of the government and industrial world's efforts to promote nuclear power generation to the Japanese people.[43]

In a way, the tapestries represent two institutionalized discourses on the nuclear. On the one hand, one discourse represented the nuclear as a terrible weapon but this had largely been relegated to the past, or projected on to monsters such as Godzilla that could be contained. The second discourse was more in favour, namely the nuclear as part of Japanese identity, a peaceful use of the atom that was safe, clean and good.[44] The Japanese had ostensibly rejected the use of nuclear weapons, despite enjoying the protection of the US nuclear umbrella, and embraced nuclear power as an integral part of its future.

Writing in 1970, art critic Haryū Ichirō noted the preponderance of images in pavilions at Expo '70 and the lack of real objects. He regarded the image-based shows that could be viewed in many pavilions as "mirages that they can't even touch."[45] He noted how many of the images were arbitrary constructs and that visitors "walked around disingenuously among these banalities that have become the official truth,"[46] consuming the official narrative about Japan's road from the past to the present prosperity. He considered the Japanese Pavilion as being where the mechanical reproduction of images was most complete. There were hardly any actual physical artefacts but rather models, photographs, graphs and films. He noted how Kōno's tapestry "Tower of Joy" served to downplay the impression of atrocity associated with the dropping of the atomic bomb.

Pavilions such as the Electric Industry Federation's Electric Power Hall emphasized the positive uses of nuclear power. Funded by nine major electric power utilities, the theme of the pavilion was "Mankind and Energy."[47] The impressive pavilion consisted of two structures. The main building was designed by the prominent Japanese architect Sakakura Junzō who had worked in Le Corbusier's atelier in Paris. The adjacent floating theatre was the work of Murata Yutaka.

Sakakura's building was a 40-metre high structure consisting of four pillars. Stretched between the pillars was a roof structure under which hung a cylindrical theatre and below that an exhibition hall some 30 metres in diameter which was referred to as the Electric Power Gallery. These two cylindrical parts of the building resembled a nuclear reactor core. Visitors would firstly enter via the hall and take an elevator up to the third floor where they would go up a spiral slope to the theatre. There they would walk past a five-screen video presentation on the cylindrical wall and then view a sixteen-minute documentary film entitled *The Sun's Story* on a large, 9-metre high, 22.5-metre wide screen. The film depicted the relationship between human beings and energy throughout the world, from the use of fire in ancient times through to the nuclear age. In this way, nuclear power was portrayed as natural and a part of human progress.[48]

Visitors would proceed down to the Electric Power Gallery where they viewed the history of electricity and how technology had evolved over time. They could go to six parts of the room where they would learn about different aspects of nuclear power. They could view an image of nuclear fission, inspect an electric power map of Japan and learn where nuclear power plants were located around the world. Particularly impressive was a large, 10-metre long glass display case which showed a 100 kV, high voltage electrical discharge.[49] Visitors could also view an actual size model of a fast breeder reactor core which conveyed the sense of being in a reactor.[50] The fun and spectacle did not end there. In the floating theatre adjacent to the main building, there were regular performances of a magic show using electricity.

Other corporate pavilions also conveyed the promise of atomic energy. The Fujipan Robot Pavilion looked at the future in a light-hearted way. Designed by the cartoonist Tezuka Osamu (1928–1989) who was appointed pavilion producer, the pavilion resembled a giant caterpillar, in keeping with its theme of "Children's Dreams." Tezuka was an Osaka-born local who had grown up on animated, Walt Disney feature films and used a simplified, animation style that owed much to Disney. His pavilion was divided into three sections: Forest of Robots, Town of Robots, and Future of Robots. The first section included an eighteenth century Japanese ***karakuri ningyō*** (mechanized doll) to suggest that there had been forerunners of modern day robots back in the Edo period. The next section, "Town of Robots" included an open plaza where visitors could view stage shows and visit an exhibition hall where visitors could play the hand-game *Janken* (rock, paper and scissors) with a robot. Near the exit of the last section, "Future of Robots," cartoons showed how robots would impact on people's lives in the years to come. In June, during the Expo, Tezuka met with the public, signed autographs and drew cartoons of some of his characters such as Tetsuwan Atomu (Mighty Atom) but more commonly referred to as "Astro Boy" outside of Japan).[51]

Whether intentionally or not, Tezuka's creation had promoted the acceptance of nuclear power from when it first appeared as a ***manga*** character in the monthly boys' magazine *Shōnen* from April 1951 to March 1952. As Yuki Tanaka has argued, Tezuka seems to have been influenced, perhaps unconsciously, by the overly optimistic visions of nuclear power that were prevalent in the 1950s.[52] In the series "Ambassador Atom," the boy robot was created by a Dr Tenma, the head of the Ministry of Science in Japan, to replace his lost son Tobio who had been killed in a traffic accident. The boy robot becomes an emissary of peace who convinces aliens to go to another planet and not invade Earth.[53] This ***manga*** story would be reworked into an animated television series that was first broadcast in Japan between 1963 and 1966. The popular television series told the story of a boy robot powered by a nuclear reactor, who can be both a cute hero and a dangerous threat to the world if his powers are not used for good.[54] Tezuka's Tetsuwan Atomu (Mighty Atom) was quickly introduced in the USA as Astro Boy, thereby removing direct reference in the character's name with nuclear power. Astro Boy has a younger sister named Uran (Uranium) but she would become known variously in the West as Astro Girl, Sarah or Zoran. The adventures of Astro Boy would show Japan how they could save the world through careful management of nuclear power.

The broadcasting of the *Astro Boy* television series coincided with a boom in toys based on the character in late 1963 and 1964. Children could surround themselves with toys and stickers of the cute, reactor-powered boy robot, make nuclear power visible in their own bedrooms.[55] In their world, nuclear power was made to seem "natural and heroic."[56] This affection for the boy robot helped to counteract Japan's so-called "nuclear allergy"[57] which was prevalent in the late 1960s and replaced it with a sense of wonderment at what science and technology could offer to the Japanese people. Astro Boy contributed to popularizing the nuclear as part of Japanese identity. He reflected the new Japan and was a symbol of Japan's science and technology-led future. His good nature was emblematic of how the Japanese sought to harness nuclear power for peaceful purposes. As Kondō Motohiro has suggested, "more than most peoples of the world, the Japanese have embraced technology without ambivalence as a partner and ally, evincing a nearly unshakable faith in its ability to improve human life."[58]

The Fujipan Pavilion at Expo '70 made the most of this by linking Tezuka and Astro Boy to a positive image of robots[59] but also to nuclear power.[60] Like the humanoid robot himself,[61] Expo '70 served to domesticate nuclear power into a friendly energy source which seemed natural and a part of everyday life. By this time, not only had the Japanese government embraced the concept of atoms for peace, it had chosen the USA as its primary partner in the development of a civilian nuclear power industry. Japanese government surveys have shown that between 1968 and 1976, the years leading up to and following Expo '70, most Japanese supported plans to continue building nuclear power plants.[62]

Expo '70 has been criticized as constituting a type of science fiction in the idealized way that it presented the theme of "progress and harmony" as if in a bubble while the nation was undergoing turmoil as a result of protests against the US-Japan Security Treaty and the Vietnam War.[63] But from the government's perspective, that served even more as justification for holding the international event in Japan, one that would distract public attention away from defence arrangements with the USA.[64]

In the wake of Expo '70, nuclear power producers built small-scale visitor centres to promote nuclear power but by the mid-1980s, large sums of money of the order of 3–10 billion yen (approx. US$15–50 million) were devoted to each centre.[65] Sumihara Noriya has traced how the average, annual number of visitors to the Fukushima Daiichi Nuclear Power Plant went from 9500 in the 1970s through to the mid-1980s but from 1985 to 1990 it grew to 37,000.[66] Sumihara has identified a common narrative at visitor centres. The rhetoric is that Japan is extremely poor in natural resources for the production of electricity and that most of its sources have to be imported. Since the oil shocks of the 1970s, it is unwise for Japan to be reliant on oil-producing nations which are in politically unstable parts of the world. The supply of oil, it is suggested, will be exhausted by the mid-twenty-first century and the use of fossil fuels contributes to global warming. Clean, alternative energy sources such as hydro-electric power still accounts for a small share of electricity production and wind and solar power are not sufficiently reliable to meet Japan's energy needs. It is then posited that nuclear power is a good alternative. What is more, nuclear fuel in the form of uranium comes from the USA and Australia, two politically stable nations that enjoy good relations with Japan. Uranium may one day be exhausted but through reprocessing, it can be used over and over again. As for safety, if the uranium is handled carefully, it does not negatively impact the environment and in terms of cost, it is the cheapest means of electricity production. The centres also reassure visitors that the power plants are designed to withstand earthquakes and typhoons. As for radioactivity, it occurs everywhere in nature.[67]

Commencement of Construction of a Calder Hall Reactor in North Korea 1980

North Korea also pursued nuclear power. With the help of the Soviet Union, the Yongbyon Nuclear Research Centre was established in 1962 as part of the first phase of the North Korean nuclear programme.[68] It included the installation of its first nuclear research reactor, a small 4MWe research reactor supplied by the Soviets that began operation in 1967.[69] In 1980, the second phase of development began with the expansion of the facilities and the start of construction of a 5MWe graphite-moderated, gas-cooled reactor. It was modelled after the British Calder Hall reactor and completed in 1986. Like the Japanese before them, the North Koreans were initially attracted to this type of dual-use reactor as a starter reactor. It used natural uranium as fuel and uranium was plentiful in North Korea. Another advantage was that the Calder Hall reactor used carbon-dioxide gas as a coolant rather than heavy water which was difficult to obtain. What is more, graphite was plentiful in North Korea.

North Korea told the International Atomic Energy Agency that the reactor would be used for civilian nuclear power but as we have learnt earlier in this book, the Calder Hall reactor could be used for the production of plutonium. North Korea saw this capability as one way of countering the threat posed by the USA which had first sent atomic-capable weapons into South Korea in 1958.[70] When North Korea announced that it had produced nuclear weapons in 2005 and conducted a test in 2006, suspicions that the reactor had been used for military purposes were confirmed.[71] It has been suggested that the reactor produced the country's entire inventory of plutonium.[72]

The Yongbyon Calder Hall reactor does not look like the original British Calder Hall reactor but does bear a similarity to the improved Calder Hall reactor at Tōkai-mura that operated from 1965 to 1998. GE Corp. (UK) had modified the design to meet Japan's seismic conditions. It appears that North Korea looked to Tōkai-mura for its own Yongbyon 5MWe reactor. As we have learnt, visitors inspected the Japan Atomic Power Company (JAPC) reactor at Tōkai-mura and three films recorded its construction. Japan Atomic Energy Research Institute (JAERI) facilities at Tōkai-mura were also open to the public and trips to see reactors were openly encouraged.

Technical information on the Calder Hall reactor at Tōkai-mura was in the public domain from early on, thanks to the Atoms for Peace program.[73] As early as March 1960, the British journal *Nuclear Power* carried articles on the engineering design, earthquake problems, optimization, control and safety as well as construction and organization of Japan's first nuclear power station.[74] Japan Atomic Power Company (JAPC) published a detailed 430-page record of its construction in late 1971.[75] It appears that North Korea took advantage of the publicly available scientific information on the Calder Hall reactor. The small Yongbyon reactor became operational in 1986[76] and continued operating until 1994. It was restarted in 2003[77] and it appears that spent nuclear fuel was reprocessed to yield plutonium metal.[78] Some of this was used for nuclear weapons in North Korean nuclear tests in 2006.[79]

In 1986, the same year that the Yongbyon reactor was completed, the well-known Japanese philosopher Umehara Takeshi was interviewed by the magazine *Genshiryoku bunka* (*Nuclear Culture*) published by the Japan Atomic Energy Relations Organization (JAERO). Although not claiming any special expertise on the topic of nuclear power, he did liken it to a god or god-send for Japan.[80] Thanks to nuclear power, a tremendous amount of energy has been made available by humans harnessing nature. Much was made of the peaceful uses of atomic energy in what was the 200th issue of the magazine. But the Chernobyl nuclear disaster had occurred in April that year and completion of the Yongbyon reactor highlighted the dual use of the Calder Hall reactor and how it could produce weapons-grade plutonium. This was the same reactor that Japan had chosen for its first nuclear power station.[81] The construction of the improved

Calder Hall reactor in both Japan and North Korea shows the arbitrary nature of the division between civilian and military uses of nuclear power and suggests mixed motivations behind its introduction in both countries.

Robert Alvarez recalled visiting the Yongbyon nuclear complex on 16 November 1994 as part of an official US delegation and he inspected the then dormant reactor. The USA and North Korea had signed an agreement on 12 October 1994 in which North Korea agreed to freeze its production of plutonium and allow spent reactor fuel to be taken out of the country. Like the Japanese before them, North Korea had come to realize that light-water reactors were better suited to producing electricity. So in return, the plan was for the USA to provide economic cooperation, fuel oil, and the construction of two light-water reactor power plants. Unfortunately, the US Congress opposed the agreement, the USA fell behind in meeting its commitments and the agreement was eventually derailed.[82]

Nevertheless, the possibility of the agreement allowed Alvarez to visit Yongbyon and view the Calder Hall reactor. When he visited the reactor room, he noted a small replica of a 1950s British nuclear power plant sitting there.[83] But the Yongbyon reactor was not a case of mimesis and simply a scaled-up version of that model. It appears that the North Koreans had looked to the improved version at Tōkai-mura which was designed to withstand earthquakes. The combination of reactor and steam-raising system there were entirely self-contained and connections with outside buildings and services could be cut without hampering reactor safety.[84] Alvarez reported how he could stand on the top of the Yongbyon reactor vessel as it had been built partially below ground level.[85] At Tōkai-mura, the main control room and other services such as carbon dioxide, reserve feed water and pumps, diesel generators and batteries were all contained within the reactor structure. In addition, a feature of the aseismic version of the reactor was that the bottom-charging of the reactor had been changed to top-charging, thereby bringing the centre of gravity closer to ground level.[86] Unfortunately, an earthquake would be the undoing of the development of nuclear power in Japan.

Plutonium Boy

We have seen how Japanese industry sought to soften the image of nuclear power at Expo '70 by embracing cute characters such as Astro Boy (aka "Mighty Atom"). The now-defunct government research organization

Power Reactor and Nuclear Fuel Development Corporation (PNC) created a new character named Puruto-kun (Plutonium Boy or Mr Pluto) who featured in a 1993 promotional video entitled *Purutoniumu monogatari: Tayoreru nakama Puruto-kun* (*Plutonium Story: Plutonium Boy, Our Reliable Friend*) that sought to debunk the fear of plutonium.[87] Some 250 copies of the video were produced and sent out to local authorities and visitor centres as part of educational outreach. The video consists of animated sections when the main character is pictured speaking to the audience, frequent hand-drawn images over which we hear the character's voice, and actual footage of nuclear facilities, production of nuclear fuel, and mundane actions such as washing hands.

The Japanese government has long seen plutonium as enabling Japan to gain energy self-sufficiency through the development of fast breeder reactors[88] and the video endorsed it. The 10.55-minute video begins with a Halloween-like gathering of ghosts and flying pumpkin heads. A somewhat friendly-looking ghost discards his white sheet to reveal cute Plutonium Boy who wears a bright green helmet clearly marked "Pu" for plutonium that extends over his ears. A pair of antennae protrude from the helmet, perhaps to signal that he emits alpha rays. He introduces himself and says that the video was produced because some people see him as being a monster, something frightening and not well understood. He explains that he regrets that plutonium was used as a tool of war as in the atomic bomb dropped on Nagasaki.[89] The character claims he hates war and loves to work for peace. He compares himself to dynamite. While it can be dangerous, it can also be used for peaceful purposes. He urges his young viewers to carefully listen to the true story of plutonium that he is about to tell. Plutonium he explains, is not mined like uranium but is created from uranium in nuclear reactors. We are told that Dr Glenn T. Seaborg who once served as the chairman of the US Atomic Energy Commission discovered plutonium in 1940. Plutonium Boy suggests that not much time has passed since that discovery and there are a lot of misconceptions about plutonium which he hoped to dispel.[90]

The first misunderstanding relates to the fear that criminals could easily build an atomic bomb if they obtain sufficient plutonium. Plutonium Boy explains that weapons-grade plutonium would have to be at least 93 per cent pure whereas that generated by reactors is only about 70 per cent pure. Also, advanced technology and large-scale equipment would be required to make an atomic bomb. Plutonium is strictly managed when transported and stored, making theft difficult. It is therefore not realistic to think that one can make an atomic bomb using the plutonium from reactors.[91]

The second misunderstanding concerning plutonium is that it can cause cancer. We are told that it is not a poison that immediately kills you when consumed like potassium cyanide. The main reason why he is called a poison is because he discharges alpha rays. Though the alpha ray is a form of radiation, it can be interrupted by just one sheet of paper. Admittedly, he does keep discharging alpha rays for a long period but even if plutonium adheres to the skin, it is not absorbed into the body. Even if one was to swallow plutonium and it enters the stomach and intestines, it is excreted out of the body. The video does acknowledge, halfway through the film, that when plutonium enters the body from a wound, it is not easy to expel. Plutonium concentrates on the lymph node and can travel to the bone and the liver, discharging alpha rays. Moreover when a person inhales plutonium, though part of it is expelled from the body, it can move to the bone and the liver via the lungs, emitting alpha rays for a long period. Thus in terms of precautions, it is important for plutonium not to enter the blood stream and not to be inhaled. So the video does acknowledge the dangers of plutonium when ingested and how humans are exposed to it can determine how harmful it is.[92]

The video claims that there have been no documented cases of someone having contracted cancer due to exposure to plutonium.[93] We are told that beliefs about the potential harm of plutonium when it enters the human body are based on data from tests on animals. This research suggests that when plutonium enters the bloodstream and lungs, and depending on the quantity and chemical form when ingested, it can cause cancer in animals. Plutonium Boy again reassures the audience that due to strict standards for treating and handling plutonium, it is not possible that a situation would arise when it would negatively impact on human beings. Even if criminals dumped plutonium into a reservoir, plutonium does not melt easily in water and most of it would sink to the bottom due to being heavy. Even if consumed in drinking water, it would not be dangerous. Plutonium Boy is shown shaking hands with a boy who drinks a glass of water tainted with plutonium. The boy subsequently goes to the toilet and the plutonium safely passes through his body, leaving him feeling great.[94]

In the last minutes of the video, Plutonium Boy suggests that people see him as a threat only when the dangerous aspects of him are emphasized. This is due to a lack of correct knowledge about plutonium and previous associations of plutonium with the atomic bomb. He hopes that the wonderful wisdom of human beings will ensure that the potential dangers will be controlled. We then are provided with an aerial view of the Monju fast breeder reactor in Fukui prefecture which is clearly cited as a manifestation of that. We are told that it is most efficient to use plutonium in a nuclear reactor. We see an image of children facing the sunrise. The use of plutonium in fast breeder reactors fills them with hope. The closing images credit the video to the Tōkai-mura office of the Power Reactor and Nuclear Fuel Development Corporation in Ibaraki prefecture that has, we are told, been conducting research and development of fast breeder reactors, advanced thermal reactors and plutonium fuel for more than twenty-five years. Plutonium Boy has the last word. He explains to everybody that he is not a monster, and urges the audience to look carefully at his true self. He hopes that people will calmly and peacefully engage with him as a friend, for he is not frightening and he is not dangerous. He is a companion who offers exhaustless energy for a long time into the future. Plutonium Boy waves goodbye, superimposed on an aerial view of nuclear facilities.

This clever down-playing of the dangers of plutonium was part of the campaign to counter public opposition to the building and 1993 start-up of the plutonium-fuelled Monju prototype fast breeder reactor by PNC in Fukui prefecture. It was hoped that construction of Monju would be followed by construction of a demonstration reactor which in turn would lead to the introduction of commercial fast breeder reactors. The nuclear power industry portrayed opposition to this strategy as being the result of a "lack of understanding" on the part of the public, but the economic rationale for using fast breeder technology to generate electricity from plutonium (rather than uranium) using "fast neutrons" to breed more plutonium than it consumes, has been much diminished since it was first considered back in 1956. Uranium is much more plentiful than originally thought.[95] PNC specialized in fast breeder reactors and had a vested interest in Monju's continued operation. It was the major promoter of fast breeder reactor development in Japan but ultimately was shut down in 1998 after a number of nuclear accidents. These accidents occurred at Monju and Tōkai-mura.

Nuclear Accidents

The Monju accident occurred during the night of 8 December 1995. The reactor had first achieved criticality in April 1994. Unlike uranium-fuelled reactors which are cooled by water, plutonium fuelled reactors are cooled with sodium. Accidents in the former can result in a melt-down whereas those at the latter can result in explosions and plutonium being released into the environment. Sodium coolant leaked from a pipe at Monju and a fire ensued. Workers could not contact anyone to obtain emergency instructions. Not only were safety procedures inadequate and the reactor construction problematic, but there was potential for people in the region to be exposed to deadly quantities of radiation. Managers supplied the media with a doctored video downplaying the extent of the sodium spill and hid a videotape that had been recorded in the room where the sodium leaked a few hours after the accident had occurred.[96] They also repeatedly lied about the accident which in reality was the worst known accident to have occurred at any fast breeder reactor in the world. Thanks to some intrepid Fukui prefectural officials who gained access to the reactor in the morning of 11 December, the world learnt about the true extent of the accident.[97] Eventually, after spending more than US$9 billion on Monju, the Japanese government decided in December 2016 to decommission the reactor. This was after twenty-two years and having only operated some 250 days during that period.[98]

The Monju accident was followed by a fire at the Tōkai-mura PNC nuclear waste reprocessing facility on 11 March 1997. Apparently, several of the maintenance staff had been absent from the facility at the time, playing golf and no fulltime plant employee was on duty. A bungled attempt to extinguish the fire resulted in an explosion that resulted in thirty-seven workers being exposed to radiation. The radiation levels turned out to have been ten times higher than had been initially reported.[99] On 17 March, the Science and Technology Agency declared it to have been Japan's worst nuclear accident and considered breaking up PNC.[100] It would be disbanded in 1998 and restructured as the Japan Nuclear Cycle Development Institute. This would in turn merge with the Japan Atomic Energy Research Institute (JAERI) in 2005 to become the Japan Atomic Energy Agency with headquarters in Tōkai-mura.[101]

The worst accident at Tōkai-mura occurred on 30 September 1999 when three workers triggered a nuclear fission chain reaction at the nuclear fuel processing plant operated by JCO Co. Ltd. (formerly known as the Japan Nuclear Fuel Conversion Co.), a subsidiary of Sumitomo Metal Mining Co. The JCO plant was one of fifteen nuclear facilities at Tōkai-mura at the time, including Japan's first commercial nuclear power plant. The workers were preparing a batch of fuel for Jōyō, an experimental fast breeder reactor that was a forerunner of Monju. Suddenly there was a flash of blue light indicating a chain reaction had begun.[102] In the morning of 1 October, after about twenty hours and a confused emergency response, the chain reaction was brought to an end. Government officials had apparently requested US military assistance but were informed that they were not equipped to deal with nuclear accidents.[103] Some 667 people including plant workers, firefighters and residents who lived adjacent to the plant were exposed to elevated levels of radiation. Two of the workers died and six managerial staff including the surviving worker who had supervised the two workers were arrested, charged with negligence and ultimately pleaded guilty.[104]

By the end of the year, a new concrete wall was erected to hide the plant from public view and a *Mainichi Daily News* poll conducted shortly after the accident found that 70 per cent of the Japanese public opposed nuclear power.[105] The entire JCO plant was shut down in 2000. JCO agreed to pay a total of US$121 million in compensation to settle 6875 cases stemming from the accident and lost it operating licence for the plant.[106] These and other accidents not only led Japanese to question the safety of nuclear power but also to ask whether Japan was losing its technological prowess,[107] a feature of the national narrative that had been so assiduously promoted since Japan's reconstruction after World War II. Indeed, even Godzilla seemed to have doubts. Tōhō released a new Godzilla sequel entitled *Godzilla 2000* in December 1999. Three major battle scenes were featured in the film, including one which had been shot on location at Tōkai-mura prior to the nuclear accident. Despite the accident, the scene was retained in the film to serve as a cautionary message about the dangers of atomic energy. In the film, energy-hungry Godzilla goes to Tōkai-mura to feed on the nuclear reactors and the Self Defence Force is mobilized to protect the facilities. Atomic energy is a double-edged sword. It helped create the mutant monster but Japan has also relied on nuclear power for up to a third of its electricity.[108]

Fukushima

As we have seen, Japan has a record of nuclear accidents and cover-ups. In September 2002, it was revealed that there had been systematic falsification of inspection and repair records at thirteen Tokyo Electric Power Company (TEPCO) reactors. TEPCO had become by this time the world's largest private electrical utility.[109] The Fukushima nuclear disaster in 2011 well and truly put paid to the myth of safety. A tsunami hit the

Fukushima Daiichi Nuclear Power Plant as a result of the magnitude 9.0, Great East Japan earthquake on 11 March that year. The tsunami flooded the TEPCO power plant, leading to a loss of emergency power which was needed to keep the reactors cool. There were core meltdowns in three of the six reactors.[110] Unfortunately, the power station had only been designed to withstand a tsunami of maximum height of 3.1 metres above mean sea level. In line with that, TEPCO positioned the seawater intake buildings at 4 metres above sea level. The main buildings were constructed at the top of a slope that was 10 metres above sea level. Just before the tsunami made landfall, it is estimated that it had a height of approximately 13.1 metres. By the time it had run up the slope, it was 14–15 metres high and reaching 17 metres in some places.[111]

As was discussed in the previous chapter, TEPCO had commissioned a documentary entitled *Reimei: Fukushima Genshiryoku Hatsudensho kensetsu kiroku, chōsa-hen* (*Dawn of a New Age: A Record of the Construction of the Fukushima Nuclear Power Plant, Preliminary Surveys*) (1967) that documented the clearing of the land that ultimately reduced the 35-metre bluff by some 25 metres, leaving the site vulnerable. Locating seawater pumps to provide coolant water from a height of 10 metres above sea level was cheaper than doing so on higher ground. In addition, TEPCO dug an additional 14 metres to create basement floors for the turbine buildings and emergency generators.[112]

On 11 March 2011, at approximately 2:46 p.m., the Daiichi Nuclear Power Plant went into automatic shutdown due to the seismic activity. When the earthquake took out external power, backup diesel generators kicked in to provide emergency power but at approximately 3:35 p.m., the tsunami struck and the water supply equipment was inundated. This included the emergency seawater system. This left the plant without any way of shedding heat and also affected the backup generators. All power was lost and this impeded management of the accident.[113]

The whole discourse about nuclear power in Japan was disrupted. Some Japanese saw what occurred at the nuclear power plant in Fukushima as the third atomic bombing of Japan after Hiroshima and Nagasaki. Whereas atomic bomb survivors had been supportive or relatively silent regarding nuclear power, they became more vocal in their opposition. As Susan Lindee has pointed out, "Fukushima-as-Hiroshima became an important trope in Japan"[114] and the nation is still grappling with the dilemma of whether to restart reactors or somehow replace nuclear power with alternative energy sources. The perception of Japanese prowess in technology, especially robotics, took a beating when it was found that the power plant at Fukushima lacked emergency robots and American robots had to be shipped in a hurry from the makers of the Roomba robotic vacuum cleaner. Perhaps the adventures of Astro Boy set unrealistic expectations among the Japanese populace regarding the capabilities of Japanese robotics. The critic Tachibana Takashi had warned of this several years earlier. There was a discrepancy between humanoid robots performing on a stage, at a special event or on television and industrial robots in the workplace or in everyday life.[115] What is more, not only did the Fukushima nuclear plant operators and nuclear regulators believe the safety myth themselves and think that an accident would not happen, but the very introduction of emergency robots prior to the disaster was considered as unnecessarily inspiring fear among workers by assuming that an accident might be possible. It has been argued that the rejection of robots was symptomatic of a reluctance to improve maintenance and invest in new technologies.[116]

In the aftermath of the disaster, *New York Times* reporter Onishi Norimitsu travelled to a visitors' centre near the Shika Nuclear Power Plant, Ishikawa prefecture, owned and operated by the Hokuriku Electric Power Co. He viewed the exhibitions which just happened to use Lewis Carroll's *Alice's Adventures in Wonderland* (1865) to portray nuclear power in a positive light. We have previously seen how Alice had featured in the promotion of atomic energy in Japan over half a century earlier in 1954. In 2011, the White Rabbit lamented to Alice how Japan was running out of energy and a Dodo robot figure recommends nuclear power to Alice as something "clean, safe and renewable if you reprocess uranium and plutonium."[117] As Alice remarked, "You could say that it's optimal for resource-poor Japan!"[118] Despite the delightful exhibitions, it was subsequently ascertained by the Nuclear Regulation Authority that the Shika Nuclear Power Plant may be situated above an active geological fault.[119] Etsuko Kinefuchi has noted how, in addition to the appropriation of Alice in Wonderland at Shika Nuclear Power Plant, Miyazaki Hayao's character of Totoro had been used at the energy museum at Fukushima Daini (No. 2) Nuclear Power Plant to persuade children that nuclear power was safe and reliable.[120]

In the aftermath of the Fukushima disaster, Japan suspended operation of all its nuclear reactors for safety inspections. Japan went from relying on nuclear power for approximately 30 per cent of its electricity to none, turning to coal, oil and natural gas for its energy needs. The Japanese government appointed a Fukushima Nuclear Accident Independent

Investigation Commission which produced a 641-page report[121] that was published in June 2012, around the time when the anti-nuclear power movement was at its peak.[122]

The report was accompanied by an English-language version of over 500 pages.[123] The Commission's findings were that "this accident was not a 'natural disaster' but clearly man-made'"[124] and the result of human negligence. In his preface to the report, the chairman, Kurokawa Kiyoshi traced the origins of the accident to the time of Japan's post-war period of high economic growth when political, bureaucratic, and business circles in Japan promoted nuclear power generation as national policy. Company and ministry elites assigned a higher priority to the interests of their respective organizations over protecting the lives of the Japanese people, delaying the implementation of necessary safety measures. He suggested that accident could be linked to the 'mindset' of the Japanese people.[125] A bureaucratic and organizational mindset is elaborated upon in the detailed report. This mindset was ill-suited to a safety culture that could absorb new knowledge and make improvements.[126] Kurokawa's preface to the Executive Summary stated it in following way:

> What must be admitted – very painfully – is that this was a disaster "Made in Japan." Its fundamental causes are to be found in the ingrained conventions of Japanese culture: our reflexive obedience; our reluctance to question authority; our devotion to 'sticking with the program'; our groupism; and our insularity.[127]

In a critique of the report, historian of science Higuchi Toshihiro, writing in the *Bulletin of the Atomic Scientists*, suggested that while culture was not irrelevant, it should not blind people from understanding that the question of safety or the lack of it was a "structural problem of nuclear regulation that can be fixed."[128] What is more, if Japanese culture and society are to shoulder the blame for the accident, they need to acknowledge that all Japanese people have become stakeholders in nuclear power. He goes on to point out how reactors had provided strength to the state, helped corporations earn profits, allowed scientists to obtain knowledge, given workers jobs, and provided a source of electricity to consumers.[129]

Shin Godzilla

The change in Japanese attitudes to nuclear power that followed the Fukushima disaster can be seen in the latest Japanese made Godzilla film. Known as *Shin Gojira*, it was released in Japan in late July 2016 and was a nationwide hit. It was released in the USA in October 2016, initially as *Godzilla Resurgence* but due to the similarity in title to the film *Independence Day: Resurgence* that was released that same year, it became known internationally as *Shin Godzilla*.[130] "Shin" can mean "new", "real" and "god".[131] The first Godzilla film to be released since the Fukushima disaster, the popularity of the film partly lies in how it depicts how the Japanese now see nuclear power in a different light. Director Anno Hideaki, creator of the anime television series "Neon Genesis Evangelion" and co-director Higuchi Shinji portray Godzilla as the result of illegal dumping of nuclear waste which caused ancient creatures to mutate.[132]

The film begins with thermal eruptions occurring in Tokyo Bay. A young, up-start bureaucrat, Yaguchi Rando correctly attributes it to a creature. This is confirmed when a crawling sea creature emerges from the waters and creates havoc in Tokyo. It quickly self-mutates into the dinosaur-like Godzilla that we know. Just like in the original 1954 version, he stomps all over Tokyo again. Some politicians and bureaucrats quickly seek the creature's extermination while scientists call for its live capture. However, this time, the film is almost comedic in how it portrays the responses of the scientists, politicians and bureaucrats, in a seeming reference to the actual incompetence of the government in dealing with the Fukushima nuclear disaster.[133] As soon as there is an official statement that there was no danger of the creature coming ashore, it duly does so and moves towards Shinagawa which is evacuated. Yaguchi dares to question the actions of his seniors but ultimately proves correct. He is quickly promoted and brings together a band of nerds and misfits to save Tokyo from devastation, not only by Godzilla but also by the USA that is keen to use nuclear weapons to kill Godzilla and contain the damage, lest Godzilla pose a threat to American shores. One of the film's messages is that Japan's bureaucratic culture must evolve with Godzilla otherwise the nation is doomed.

There is a soft nationalism apparent in the film and some criticism of the US-Japan relationship. Yaguchi's team enlists the help of the Self-Defence Forces that are portrayed in a positive light. This reflects how many Japanese applauded the speedy and large-scale deployment of troops for the search and rescue of victims of the Great East Japan earthquake. In the film, the Japanese fall victim to Godzilla who resurfaces in Sagami Bay, nearly doubled in size. It is attacked and fired at with missiles but to no avail. Meanwhile US Air Force bombers are on their way with nuclear weapons. The prime minister declares that he doesn't want to go down in history as sanctioning the third nuclear bombing of Japan.[134] The film very briefly shows images of Hiroshima and Nagasaki and there is a comment that "post-war Japan is a tributary state" of the USA. An ambitious Japanese-American special envoy named Kayoko Ann Patterson, the daughter of a US senator, is sympathetic to the plight of the Japanese people and is determined not to allow her fellow Americans to be responsible for a "third mistake". People are fearful that Tokyo would end up as a "kikan konnan kuiki" ("difficult to return to zone"), a euphemism used to refer to the contaminated zone around Fukushima.[135]

Godzilla breathes fire and emits destructive rays from its back, leaving high levels of radiation in its wake. Yaguchi's team determines that radioactive Godzilla's internal organs are like a nuclear reactor and the circulation of blood distributes the heat that is generated. There is a nationwide, industrial effort to produce a coagulant to stop Godzilla. Meanwhile, the French ambassador buys the Japanese some time by temporarily delaying the UN Security Council-sanctioned bombing of Tokyo. Bomb-laden, unmanned bullet trains crash into Godzilla and buildings are exploded, collapsing on top of the creature. While Godzilla is down, the Self-Defence Force administers a blood clotting agent through Godzilla's mouth to freeze and immobilize him. But Godzilla is not dead. The threat of American nuclear weapons being used on Godzilla and Tokyo remains as does the issue of the dumped nuclear waste. Some of the general sentiment expressed in the film resonates with the findings of a public opinion survey on nuclear power conducted the year after the film was released.

Changed Attitudes to Nuclear Power

In October 2017, the Japan Atomic Energy Relations Organization (JAERO) surveyed 1200 Japanese aged between fifteen and seventy-nine years of age by paying a personal visit to their homes. Not surprisingly, negative images of nuclear power were greater than positive images. A tiny 0.9 per cent of respondents thought that it was good and 19.1 per cent thought it was bad. Undoubtedly linked to that was a perception that nuclear power was dangerous (68.5 per cent) and unreliable (30.2 per cent) in contrast to being safe (1.8 per cent) and reliable (0.8 per cent). JAERO did find, however, that more people thought that it was beneficial (17.8per cent) than useless (2.0 per cent) and necessary (17.9 per cent) compared to unnecessary (13.6 per cent). There was also the general finding that the public image of nuclear power tends to fluctuate in the wake of an accident and that experts, power utilities and local and national levels of government were not to be trusted when it came to nuclear power. Most respondents felt that it was unavoidable that they had to use nuclear power for some time but that it should be phased out.[136]

Since 2015, a small number of reactors have been restarted and reconnected with the electricity grid. In FY2017, five were in operation in Japan.[137] But despite the resumption of nuclear power, it is noteworthy that even the Ministry of Economy, Trade and Industry in its draft *Fifth Energy Basic Plan* released in mid-2018, openly acknowledged that

> nuclear operators must continue to reflect on the fact that they fell into the trap of the so-called "myth of safety," resulting in the failure to adequately deal with the severe accident [at Fukushima] and prevent a disaster like this.[138]

The plan nevertheless has an unrealistic target of nuclear power accounting for 20 per cent of power generation in FY2030 and an optimistic figure of 30 reactors operating.[139]

To ease concerns regarding contamination and what had taken place at Fukushima, TEPCO has launched an online virtual tour of the crippled power plant in Japanese and English so that visitors can see the progress being made on the decommissioning of the facilities. Visitors can venture to the red zone where full protective clothing is necessary, from the safety and comfort of their home but they will also discover that much of the plant is in the green zone where regular uniforms are sufficient and special precautions are not necessary. They will also see that not all the reactors were affected. The No. 5 Unit that was located on higher ground was unscathed by the tsunami.[140]

As of 2018, tourists can pay 23,000 yen (approx. US$208.75) to visit the Fukushima Daiichi nuclear plant as part of a day trip from Tokyo. Local residents see nuclear tourism as a way to promote their towns and to ease radiation fears among the general public. Fukushima prefecture plans to encourage tourism by building a memorial park and archival centre complete with video displays and exhibits documenting the triple disaster of the earthquake, tsunami and nuclear meltdown that occurred there. The feeling is that if visitors see the impact of the disaster for themselves, they will come to understand that Japan needs to prevent it from happening again.[141] Such tours are also important in showing how much progress has been made in the reconstruction effort. In February 2019, a nationwide poll of 2000 people aged eighteen years or over was conducted by Jiji Press. The poll found that 42.8 per cent of the sample believed some progress had been made in the reconstruction of disaster-hit areas and 43.8 per cent did not see much progress. The majority of those polled (74.8 per cent) saw little or no progress in the areas directly damaged by the nuclear meltdown.[142]

Tokyo 2020 Olympics

The introduction of civilian nuclear power in Japan provided the Japanese people with a nationwide programme on which to build their hopes for the future. Little would they know that it would be accompanied by suffering and that this would also serve to unify the Japanese people. The *Tokyo 2020 Guidebook* (2017) for the games promoted the Tokyo 2020 Olympics as the "Recovery and Reconstruction Games."[143] To this end, the torch relay was scheduled to begin on 26 March 2020 in Fukushima prefecture with the games held in Tokyo from 24 July to 9 August 2020. The games have now been postponed to July 2021.

In the same way that Sakai Yoshinori was final torch bearer for the 1964 Tokyo Olympics, the torch relay will still be heavy in symbolic meaning. It will acknowledge the region's recovery, show to the world that the people of the affected region are indeed alive, and enable Japan to thank the world for its assistance.[144] The Fukushima-born Minister for Reconstruction Yoshino Masayoshi saw this as a way of marking the Olympics as the Games of Recovery. He hoped that survivors of the disaster would participate in the relay as torch bearers.[145]

After being lit in Greece, the Olympic flame was transported to Japan. On arrival in Japan, the flame was to have been displayed during the period 20–25 March 2020 for two days each in the areas most affected by the 2011 earthquake and tsunami: Miyagi, Iwate and Fukushima prefectures. The so-called "Flame of Recovery" was then scheduled to travel to all forty-seven prefectures with the underlying message to the Japanese people that "Hope Lights Our Way." The torch relay is seen as an opportunity for the Japanese people to show how they support, accept and encourage each other.[146] In June 2019, torchbearer applications were called from those who had contributed to their respective local communities by providing support for family, friends and colleagues, as well as those who had overcome great adversity. Preference was given to those who were accepting of people from different backgrounds, welcomed newcomers into their communities and promoted harmony.[147]

The torch itself was specially designed to resonate with the symbolism of the games and the timing of the torch relay itself, during cherry blossom season. Seen from the top, the torch resembles a cherry blossom, Japan's most loved flower. Flames will be generated from each of the five petals and come together at the centre of the torch which will be made using aluminium extrusion technology, the same technology used in Japan's bullet trains. In turn, the metal itself will be recycled from aluminium waste from temporary housing that was used to shelter survivors of the earthquake. To promote the torch relay, a graphic design consisting of three rectangular shaped vermilion flames will appear to be emanating from an ochre-shaped block of colour. To achieve this effect, the "*fukibo-kashi*" method of gradated colour found in traditional Japanese woodblock prints will be used to suggest the heat and movement of the flames. In all these ways, the torch is seen as embodying aspects of Japanese culture.[148] The Olympic flame will also feature in the opening and closing ceremonies which will reflect the four themes of peace, coexistence, reconstruction, and the future, with rebirth of the affected region being an area of emphasis.[149]

Osaka Expo 2025

Not only have Godzilla and the summer Olympics returned to Tokyo but in 2018 it was announced that Osaka had been awarded the right to host the 2025 World Expo. Osaka governor Matsui Ichiro hoped to exceed the legacy of the 1970 Expo. Expo 2025 would fittingly take place on Yumeshima (Dream Island), a man-made island in Osaka Bay. In addition, there would be satellite areas in nearby Kyoto and Kobe. Reflecting this, the official name of the event would be the Osaka-Kansai Japan Expo. It would be held between 3 May and 3 November.[150]

The Expo theme will be "Designing Future Society for Our Lives" and focus on lives in three areas: saving lives, empowering lives and connecting lives. This new rhetoric speaks to the recognition that Japan must better meet the needs of the individuals who make up the nation. Expo 2025, like Expo'70 before it, would emphasize futuristic technologies, but this time in the fields of medicine and life sciences with special emphasis on wellness and meeting the needs of an ageing society. There is also an element of nostalgia for the past. Echoing the role of the Nobel Prize-winning physicist Yukawa Hideki in the initial promotion of nuclear power in Japan, co-winner of the 2012 Nobel Prize in Physiology or Medicine, Yamanaka Shinya was a key speaker at Osaka's presentation in Paris to secure the Expo. He related how it was when he visited the 1970 Expo as an eight-year-old boy that he experienced the wonders of life science. He hoped to help make Expo 2025 into a great laboratory which highlighted the beauty of life, and enthral and amaze future scientists from all over the world in the same way that had happened to him.[151]

The Osaka prefectural government hopes to attract those, like Yamanaka, who may have attended Expo '70 as a child but who now have retired and face other challenges. It was hoped that Expo 2025 would showcase advances in the health industry and include displays of robots for elderly care.[152] What hasn't perhaps changed is that Japan continues to look to science and technology for its visions for the future, this time for a greyer population that is getting smaller. It is counting on nostalgia for a time when Japan was enjoying rapid economic growth linked to large-scale, national projects organized by the state and fuelled by the "extremely obsolete development-oriented ideology of postwar Japan."[153]

Conclusion

This book has demonstrated the crucial role of representations in shaping narratives not only about nuclear power but also about the Japanese people themselves, their past and future, and the role of science and technology and the US-Japan relationship in their lives. How the Japanese people view nuclear power has been shaped by these discourses. Major events such as the Olympic Games and international expositions could and will help shape Japan's narrative going into the future. But as the Korean use of a Calder Hall reactor similar to Japan's showed us, visions of nuclear power in Japan have been blinkered. Reactors could be used to produce plutonium used in nuclear weapons.

Richard J. Samuels has found, in the wake of the Fukushima disaster, a master narrative is still under construction. People framed the catastrophe in terms of how it proved that they had been right about things. Tokyo Electric Power Company (TEPCO), the operator of the Fukushima

Daiichi Nuclear Power Plant along with government nuclear safety regulators were singled out for criticism.[154] As mentioned above, even METI where the Nuclear and Industrial Safety Agency had been located, acknowledged the myth of safety that had been pervasive. The Agency was ultimately abolished and replaced in 2012 by a more independent Nuclear Regulation Authority (NRA) that operates as an external organ of the Ministry of Environment.[155]

As the film *Shin Godzilla* shows us, people have rallied behind the SDF and ideas of military heroism. The Japanese whom we see in the film defined themselves as a community in the face of the threat posed by Godzilla (read "nuclear power") and in the process highlighting how their interests are distinct from their American friends. This resonates with the words of Ernest Ronan when he delivered a lecture at the Sorbonne in Paris on 11 March 1882. He said that "A nation is …a large-scale solidarity, constituted by the feeling of the sacrifices that one has made in the past and of those that one is prepared to make in the future."[156]

Today, you can still take a trip to the reactor, only this time there is no need to take a train and a bus, visit a department store or enter a museum to view an exhibition or film. You don't need to go to a special place anymore. Media platforms are ubiquitous and with a smart phone, one take a virtual tour without the fear associated with any real visit. There is a danger that the Japanese people will be lulled into a false sense of safety again by such easy access to images of reactors but there is growing awareness that the Japanese people are no longer spectators viewing nuclear power from afar from controlled environments. The people of Fukushima are active participants, recovering from an ongoing ordeal that is very real. They have, as Tessa Morris-Suzuki has pointed out, grown more sceptical about official scientific pronouncements provided by the government and international organizations as the information often does not accord with their everyday experiences.[157]

However, as the 2017 JAERO survey found, many Japanese feel that the use of nuclear power may be unavoidable. Despite their resignation to a future that included nuclear power, there are still concerns about how realistic government targets are. The Japan Atomic Industrial Forum conducted its annual survey of the nuclear industry and it found that half of its respondents thought that the target of nuclear power making up 20–22 per cent of Japan's total energy mix by 2030 was not possible. Even the industry itself thinks such figures are overly optimistic. There was a perception that not only did the government need to consistently promote nuclear policy and restart nuclear power plants but it also needs to restore public confidence in nuclear power.[158]

After the nuclear meltdowns at three reactors at the Fukushima Daiichi Nuclear Power Plant, many Japanese no longer believe the mass media discourses that they have been so heavily exposed to. Some Japanese have sought to bring about change. In the next chapter, we consider what we have learnt in this book and how it has informed forms of visual activism that are arguably features of Japanese protest culture.

Notes

1. Robert S. Norris, "U.S. Weapons Secrets Revealed," *Bulletin of the Atomic Scientists* 49, no. 2 (Mar. 1993): 48; Jesse Johnson, "In First, U.S. Admits Nuclear Weapons Were Stored in Okinawa during Cold War," *Japan Times*, 20 Feb. 2016, https://www.japantimes.co.jp/news/2016/02/20/national/history/first-u-s-admits-nuclear-weapons-stored-okinawa-cold-war/#.XOYjg-gzaUk; William Burr, Barbara Elias and Robert Wampler, eds, "Nuclear Weapons on Okinawa Declassified December 2015, Photos Available since 1990," *Briefing Book* no. 541 (Washington D.C.: National Security Archive, George Washington University, 19 Feb. 2016), https://nsarchive.gwu.edu/briefing-book/japan-nuclear-vault/2016-02-19/nuclear-weapons-okinawa-declassified-december-2015
2. Toshio Miyake, "Popularising the Nuclear: Mangaesque Convergence in Postwar Japan," in *Rethinking Nature in Contemporary Japan: Science, Economics, Politics*, eds. Marcella Mariotti, Toshio Miyake and Andrea Revelant (Venice: Edizioni Ca' Foscari, 2014), 71–93, esp. 72.
3. "Tokyo Gets '64 Olympic Games," *Japan Times*, 27 May, 1959, 1.
4. "Gov. Azuma Pledges Full Support," *Japan Times*, 27 May, 1959, 1.
5. Sakai Yoshinori, "Wakai sedai no eiyo" ("Young Generation's Honour"), *Asahi Shimbun*, 19 Aug. 1964, 15.
6. Jeremy Walker, "Sakai Was Fired Up for '64 Tokyo Games," *Asahi Shimbun*, 16 Jul. 2004, Asahi Kikuzo II Visual database.
7. "Kangeki no 'Kamoshika kun'" ("Deeply Moved Young Man Nicknamed 'Japanese Antelope'"), *Asahi Shimbun*, 11 Aug. 1964, evening edition, 6.
8. *Sports Illustrated*, Oct. 19, 1964, cover.
9. "Seika saishū rannā" ("Final Torchbearer"), *Shūkan shinchō* (*Weekly Shinchō*), 14–21 Aug. 2008, 62.
10. Sakai, "Wakai sedai no eiyo."
11. "Boy Born on Day A-Bomb Fell Chosen to Light Olympic Flame," *New York Times*, 23 Aug. 1964, 8.

12. "Boy Born on Day A-Bomb Fell."
13. Sakai, "Wakai sedai no eiyo."
14. "Kangeki no 'Kamoshika kun'."
15. "Shinku no honō" ("Hot Crimson Flame"), *Asahi Shimbun*, 10 Oct. 1964, evening edition, 11.
16. Uri Friedman, "Hiroshima and the Politics of Apologizing," *The Atlantic* 26 May, 2016, https://www.theatlantic.com/international/archive/2016/05/obama-hiroshima-apology-nuclear/483617/
17. Yoshikuni Igarashi, "The Bomb, Hirohito, and History: The Foundational Narrative of United States-Japan Postwar Relations," *Positions* 6, no. 2 (Fall 1998): 261–302, esp. 288.
18. "Boy Born on Day A-Bomb Fell."
19. Kuwajima Kisha (Reporter Kuwajima), "Taiyaku o hatashita Sakai-kun" ("Young Sakai Who Played a Big Role"), *Asahi Shimbun*, 11 Oct. 1964, 14.
20. "Kangeki no `Kamoshika kun'."
21. Paul Droubie, "Phoenix Arisen: Japan as Peaceful Internationalist at the 1964 Tokyo Summer Olympics," *International Journal of the History of Sport* 28, no. 16 (Nov. 2011): 2309–2322.
22. Walker, "Sakai Was Fired Up for '64 Tokyo Games."
23. "Boy Born on Day A-Bomb Fell."
24. Fintan Hoey, *Satō, America and the Cold War: U.S.-Japanese Relations, 1964–72* (Houndmills, Basingstoke: Palgrave Macmillan, 2015), 13–14.
25. Hiroshi Hamota, "Olympics Flame Sent Message to World About New Postwar Japan," *Asahi Shimbun*, 31 Mar. 2003, Asahi Kikuzo II Visual database.
26. "Japanese Protests Fade as Atom Submarine Leaves," *New York Times*, 14 Nov. 1964, 2.
27. Edwin Reischauer, U.S. Embassy Japan to U.S. Department of State, secret telegram, Nov. 14, 1964, 2, *Digital National Security Archive Collection: Japan and the U.S., 1960–1976.*
28. Reischauer, secret telegram, 2.
29. George R. Packard, "They Were Born When the Bomb Dropped," *New York Times Magazine*, 29 Aug. 1965, 28, 92–95, 98–100, esp. 99.
30. Packard, "They Were Born When the Bomb Dropped."
31. Japan Travel Bureau, *Your Technical Tour in Japan* (Tokyo: Nihon Keizai Shimbun, 1965).
32. Seiichi Takahata, *Industrial Japan and Industrious Japanese* (Osaka: Nissho Co., 1968), 144.
33. "World: Toward the Japanese Century," *Time* (2 Mar. 1970), http://content.time.com/time/magazine/article/0,9171,904215,00.html
34. Yoshimi Shunya, *Yume no genshiryoku* (*Atoms for Dreams*) (Tokyo: Chikuma Shobō, 2012), 15; Takehiro Hashizume, "Kisha to gen-

shiryoku" ("Atomic Energy As Seen by a Journalist"), *Nihon Genshiryoku Gakkaishi* 41, no. 2 (1999), 86–87; Kōichi Shikama, "'Banpaku ni genshi no akari o', hairo de omoidashita Mihama shuzai" ("'Atomic Light for the Expo' and Coverage at the Decommissioned Nuclear Reactor at Mihama," *Sankei Shimbun*, 19 Mar. 2015, http://www.sankei.com/west/news/150319/wst1503190023-n1.html; "Nuclear Power in Fukui," Fukui Prefectural Environmental Radiation Research and Monitoring Centre,, accessed 7 Feb. 2019, http://www.houshasen.tsuruga.fukui.jp/en/pages/radiation/plant/plant2.html

35. Samuel Jameson, "Japan Top Expo Attraction," *Chicago Tribune*, 22 Mar. 1970, 18, http://archives.chicagotribune.com/1970/03/22/page/46/article/japan-top-expo-attraction
36. See "Takashi Kono's Profile," Gallery 5610, http://www.deska.jp/profile, accessed 28 Apr. 2015. See also "Takashi Kono," *Idea No. 60*, Tenth Commemorative Special Issue (August 1963), accessed 28 April 2015, http://www.idea-mag.com/en/publication/060.php; "Graphic Design of Kono Takashi," National Museum of Modern Art, Tokyo, accessed 28 Apr. 2015 http://www.momat.go.jp/CG/KONO/
37. "Ōsaka banpaku Nihon kan ni tenji no tapesutorii 36 nenburi ni kōkai" ("Osaka Expo Japan Pavilion's Tapestries on Public Display Again after 36 Years"), http://www.asahi.com/culture/news_culture/OSK200609080052.html, 8 Sept. 2006; *Official Report of the Japan World Exposition, Osaka, 1970*, Vol. 1 (Osaka: Commemorative Association for the Japan World Exposition (1970), 1972), 180.
38. Japanese Government, *Nihon to Nihonjin: Nihonkan* (*Japan and the Japanese: The Japan Pavilion*) (Tokyo: Ministry of International Trade and Industry and the Japan External Trade Organization, 1970).
39. *Official Report of the Japan World Exposition, Osaka, 1970*, Vol. 1, 182–84.
40. *Japan and the Japanese: The Japan Pavilion*, English language version (Tokyo: Ministry of International Trade and Industry and the Japan External Trade Organization, 1970), 23.
41. Moritaki Ichirō, *Kaku to jinrui wa kyōzon dekinai: Kaku zettai hitei e no ayumi* (*Nuclear Weapons and Human Being Cannot Co-Exist: Steps Towards the Absolute Negation of Nuclear Weapons*) (Tokyo: Nanatsumori Shokan, 2015), 20–21.
42. Japanese Government, *Nihon to Nihonjin: Nihonkan*, esp. 23.
43. Moritaki, *Kaku to jinrui wa kyōzon dekinai*, 21.
44. Miyake, "Popularising the Nuclear," 73.
45. Haryū, Ichirō, trans. by Ignacio Adriasola, "Expo '70 as the Ruins of Culture (1970)," *Review of Japanese Culture and Society* 23, Expo '70 and Japanese Art: Dissonant Voices (Dec. 2011): 44–56, esp. 52.
46. Haryū, "Expo '70," 52.

47. Yoshikazu Miyazaki, "The Japanese-Type Structure of Big Business," in *Industry and Business in Japan*, ed. Kazuo Sato (White Plains, New York: M.E. Sharpe, 1980), 285–343 esp. 293–296.
48. "Denryokukan" ("Electric Power Pavilion"), *Expo '70 Commemorative Park*, accessed 31 May, 2019, https://www.expo70-park.jp/cause/expo/electric_power/
49. "Denryokukan."
50. Jōmaru, Yōichi, *Genpatsu to media: Shimbun jyānarizumu nidome no haiboku* (*Nuclear Power Generation and the Media: The Second Failing of Newspaper Journalism*) (Tokyo: Asahi Shimbun Publications, 2012), 200.
51. *Official Report of the Japan World Exposition, Osaka, 1970*, Vol. 1, 454–55.
52. Yuki Tanaka, "War and Peace in the Art of Tezuka Osamu: The Humanism of His Epic Manga," *The Asia-Pacific Journal* 8, issue 38, no. 1 (20 Sept. 2010): 1–15.
53. Frederik Schodt, *The Astro Boy Essays: Osamu Tezuka, Mighty Atom, and the Manga/Anime Revolution* (New York: Stone Bridge Press, 2007), 16–17.
54. Alicia Gibson, "Out of Death, an Atomic Consecration to Life: *Astro Boy* and Hiroshima's Long Shadow," *Mechademia* 8, Tezuka's Manga Life (2013), 313–20.
55. Marc Steinberg, "Anytime, Anywhere: *Tetsuwan Atomu* Stickers and the Emergence of Character Merchandizing," *Theory, Culture & Society* 26 (2–3): 113–138, esp. 125–126, 132, note 18.
56. Noriko Manabe, "Monju-kun: Children's Culture as Protest," in *Child's Play: Multi-Sensory Histories of Children and Childhood in Japan*, eds. Sabine Frühstück and Anne Walthall (Berkeley: University of California Press, 2017), 264–285, esp. 266.
57. Glenn D. Hook, "The Nuclearization of Language: Nuclear Allergy as Political Metaphor," *Journal of Peace Research* 21, no. 3 (1984): 259–75.
58. Kondō, Motohiro, "Japanese Creativity: Robots and *Anime*," *Japan Echo* 30, no. 4 (Aug. 2003): 6–8, esp. 7.
59. Gunhild Borggreen, "Ruins of the Future: Yanobe Kenji Revisits Expo '70," *Performance Paradigm*, 2 (March 2006): 123–136.
60. Schodt, *The Astro Boy Essays*, 130–31.
61. Etsuko Kinefuchi, "Nuclear Power for Good: Articulations in Japan's Nuclear Power Hegemony," *Communication, Culture & Critique* 8 (2015): 448–465.
62. Daniel P. Aldrich, "Rethinking Civil Society-State Relations in Japan after the Fukushima Accident," *Polity* 45, no. 2 (Apr. 2013): 249–264, esp. 255.
63. William O. Gardner, "The 1970 Osaka Expo and/as Science Fiction," *Review of Japanese Culture and Society* 28 (Dec. 2011): 26–43, esp. 26.

64. Richard Storry, "Defence and Economic Links with America," *Financial Times*, 29 Nov. 1967, 17.
65. Noriya Sumihara, "Flamboyant Representation of Nuclear Power Station Visitor Centers in Japan: Revealing or Concealing, Or Concealing by Revealing?" *Agora: Journal of International Center for Regional Studies* 1 (2003): 11–29.
66. Sumihara, "Flamboyant Representation."
67. Sumihara, "Flamboyant Representation."
68. Joseph S. Bermudez Jr., *North Korea's Development of a Nuclear Weapons Strategy*, North Korea's Nuclear Futures Series (Washington, D.C.: US-Korea Institute at SAIS, Johns Hopkins University, 2015), 9–10.
69. US CIA, Directorate of Intelligence, "North Korea's Expanding Nuclear Efforts," secret intelligence appraisal, 3 May, 1988, Digital National Security Archive collection: Korea, 1969–2000; Walter C. Clemens Jr., "Korea's Quest for Nuclear Weapons: New Historical Evidence," *Journal of East Asian Studies* 10, no. 1 (Jan.-Apr. 2010): 127–154, esp. 131.
70. Walter Pincus, "The Dirty Secret of American Nuclear Arms in Korea," *New York Times*, 19 Mar. 2018, https://www.nytimes.com/2018/03/19/opinion/korea-nuclear-arms-america.html
71. "Yongbyon 5MWe Reactor," Nuclear Threat Initiative, 19 July 2018, https://www.nti.org/learn/facilities/766/
72. Siegfried S. Hecker, Chaim Braun and Chris Lawrence, "North Korea's Stockpiles of Fissile Material," *Korea Observer* 47, no. 4 (Winter 2016): 721–749, esp. 723.
73. Robert Alvarez, "No Bygones at Yongbyon," *Bulletin of the Atomic Scientists* 59, no. 4 (Jul.-Aug. 2003): 38–45, esp. 41; Joshua Pollock, "Why Does North Korea Have a Gas-Graphite Reactor?" *Arms Control Wonk*, 16 Oct. 2009, https://www.armscontrolwonk.com/archive/502504/why-does-north-korea-have-a-gas-graphite-reactor/
74. See P.A. Lindley and K.J. Mitchell, "Engineering Design," *Nuclear Power* (March 1960): 104–107.
75. Watanabe Ichirō, ed., *Tokai Hatsudensho no kensetsu: Genshiryoku hatsuden paionia no kiroku* (*The Construction of the Tokai Power Station: A Record of a Nuclear Power Pioneer*) Tokyo: Japan Atomic Power Company, 1971.
76. Siegfried S. Hecker, Sean C. Lee and Chaim Braun, "North Korea's Choice: Bombs over Electricity," *The Bridge: Linking Engineering and Society* (Summer 2010): 5–12, esp. 6.
77. David Lowry, "What Theresa May Forgot: North Korea Used British Technology to Build Its Nuclear Bombs," *Ecologist: The Journal for the Post-Industrial Age*, 26 Jul. 2016, https://theecologist.org/2016/jul/26/what-theresa-may-forgot-north-korea-used-british-technology-build-its-nuclear-bombs

78. Siegfried S. Hecker, "Report on North Korean Nuclear Program," Center for International Security and Cooperation, Stanford University, 15 Nov. 2006, https://cisac.fsi.stanford.edu/publications/report_on_north_korean_nuclear_program
79. Thom Shanker and David E. Sanger, "North Korean Fuel Identified as Plutonium," *New York Times*, 17 Oct. 2006, A11.
80. Umehara Takeshi, "Genshiryoku wa Nihon no kami ni niteiru: Kōdo na ningen no chie de chinkon o" ("Nuclear Power Is Like a God to Japan: Advanced Knowledge Brings About the Repose of Souls"), *Genshiryoku bunka* (*Nuclear Culture*) 17, no. 8 (August 1986): 3–8.
81. Hecker et al., "North Korea's Choice."
82. Hecker et al., "North Korea's Choice," 7–8.
83. Alvarez, "No Bygones at Yongbyon," 41.
84. Lindley and Mitchell, "Engineering Design."
85. Alvarez, "No Bygones at Yongbyon," 41.
86. Lindley and Mitchell, "Engineering Design."
87. Power Reactor and Nuclear Fuel Development Corporation, *Purutoniumu monogatari: Tayoreru nakama Puruto-kun* (*Plutonium Story: Plutonium Boy, Our Reliable Friend*) (1993), uploaded as "Japanese Nuclear Propaganda Cartoon," https://www.youtube.com/watch?v=Iw1LYthC4PQ&t=8s, viewed 10 Jan. 2020.
88. For a detailed history, see Hitoshi Yoshioka, "The Rise and Fall of Fast Breeder Development in Japan," *Science Studies* 9, no. 2 (1996): 14–26.
89. Power Reactor and Nuclear Fuel Development Corporation, *Purutoniumu monogatari*.
90. Power Reactor and Nuclear Fuel Development Corporation, *Purutoniumu monogatari*.
91. Power Reactor and Nuclear Fuel Development Corporation, *Purutoniumu monogatari*.
92. Power Reactor and Nuclear Fuel Development Corporation, *Purutoniumu monogatari*.
93. Peter Hadfield, "Let's Have a Cup of Nice, Safe Plutonium," *New Scientist*, no. 1910 (29 Jan. 1994), https://www.newscientist.com/article/mg14119101-400-lets-have-a-cup-of-nice-safe-plutonium/
94. Chester Dawson, "The Lighter Side of Plutonium," *WSJ Blogs*, 29 Mar. 2011, https://blogs.wsj.com/japanrealtime/2011/03/29/the-lighter-side-of-plutonium/
95. Hadfield, "Let's Have a Cup of Nice, Safe Plutonium;" Masa Takubo, "Closing Japan's Monju Fast Breeder Reactor: The Possible Implications," *Bulletin of the Atomic Scientists* 73, no. 3 (2017): 182–187, esp. 186.
96. David Swinbanks, "Sodium Leak Blots Japan's Nuclear Prospects," *Nature* 379, no. 6562 (18 Jan. 1996): 196.

97. Aileen Mioko Smith, "Nuclear Mayhem at Monju," *Earth Island Journal* 11, no. 2 (Spring 1996): 22.
98. Takubo, "Closing Japan's Monju Fast Breeder Reactor," 182.
99. Peter Aldhous and Zena Iovino, "Japan's Record of Nuclear Cover-Ups," *New Scientist* 209, no. 2805 (26 March, 2011): 11.
100. "Follow-up Tokai-Mura and Leak at Fugen ATR," *Nuclear Monitor Issue*, no. 471 (25 Apr. 1997), https://www.wiseinternational.org/nuclear-monitor/471/follow-tokai-mura-and-leak-fugen-atr
101. "Political Meltdown," *The Economist* (17 Apr. 1997), https://www.economist.com/asia/1997/04/17/political-meltdown
102. For details, see Barbara Gross Levi, "What happened at Tokaimura?" *Physics Today* 52, no. 12 (Dec. 1999): 52–54.
103. Howard W. French, "Reaction Finally Controlled: 35 Exposed," *New York Times*, 1 Oct. 1999, A1, A10.
104. Michael E. Ryan, "The Tokaimura Nuclear Accident: A Tragedy of Human Errors," *Journal of College Science Teaching* 31, no. 1 (Sept. 2001): 42–48; Aldhous and Iovino, "Japan's Record of Nuclear Cover-Ups," 11; Toni Feder, "Japan Arrests Six in Nuclear Accident that Killed Two," *Physics Today* 53, no. 12 (2000): 61–62; Sandy Smith, "JCO Employees Plead Guilty to Negligence in Deaths at Japanese Nuclear Facility," *Safety Online*, 24 Apr. 2001, https://www.safetyonline.com/doc/jco-employees-plead-guilty-to-negligence-in-d-0001
105. Jean Kumagai, "In the Wake of Tokaimura, Japan Rethinks Its Nuclear Future," *Physics Today* 52, no. 12 (Dec. 1999): 51–52, esp. 52.
106. Ryan, "The Tokaimura Nuclear Accident;" Gross Levi, "What happened at Tokaimura?"; Smith, "JCO Employees Plead Guilty."
107. Calvin Sims, "Angst at Japan Inc.: A Nation Frets Over a String of Technological Accidents," *New York Times*, 3 Dec. 1999, C1.
108. Calvin Sims, "A Curtain Call for Godzilla, Back from the Dead (Again)," *New York Times*, 2 Dec. 1999, E1-E2.
109. Howard W. French, "Safety Problems at Japanese Reactors Begin to Erode Public's Faith in Nuclear Power," *New York Times*, 16 Sept. 2002, A10.
110. "Learning from the Lessons of 3/11, Seven Years On," *Japan Times*, 10 Mar. 2018, Dow Jones Factiva.
111. James M. Acton and Mark Hibbs, *Why Fukushima Was Preventable*, The Carnegie Papers (Washington, DC: Carnegie Endowment for International Peace, 2012), 9–10.
112. Reiji Yoshida and Takahiro Fukada, "Fukushima Plant Site Originally was a Hill Safe from Tsunami," *Japan Times*, 13 Jul. 2011, Dow Jones Factiva.

113. Masayoshi Matsuura, Kohei Hisamochi, Shinichiro Sato and Kumiaki Moriya, "Lessons Learned from Fukushima Daiichi Nuclear Power Station Accident and Consequent Safety Improvements," *Hitachi Review* 62, no. 1 (2013): 75–80, esp. 75–76.
114. Susan Lindee, "Survivors and Scientists: Hiroshima, Fukushima, and the Radiation Effects Research Foundation, 1975–2014," *Social Studies of Science* 46, no. 2 (2016): 184–209, esp. 199.
115. Tachibana, Takashi, "'Robotto ōkoku' Nihon no mōten" ("The Blind Spot of Robot Kingdom Japan"), *Bungei shunju* 81, no. 6 (May 2003): 144–159.
116. Norimitsu Onishi, "'Safety Myth' Left Japan Ripe for Nuclear Crisis," *New York Times*, 25 Jun. 2011, A1, A6, esp. A1.
117. Onishi, "'Safety Myth'," A1.
118. Onishi, "'Safety Myth'," A1.
119. "Fault under Shika Nuclear Reactor Likely to Be Active: NRA Expert Panel," *The Mainichi*, 3 March 2016, accessed 31 Jan. 2019, https://mainichi.jp/english/articles/20160303/p2a/00m/0na/015000c; "Faults under Japan's Shika NPP Said to Be Active," *Nuclear Engineering International*, 4 Mar. 2016, accessed 31 Jan. 2019 https://www.neimagazine.com/news/newsfaults-under-japans-shika-npp-said-to-be-active-4830171
120. Kinefuchi, "Nuclear Power for Good," esp. 454.
121. National Diet of Japan, Fukushima Nuclear Accident Independent Investigation Commission, *Hōkokusho* (*Report*) (Tokyo: The National Diet, 2012).
122. Alexander Brown, "The Anti-Nuclear Movement and Street Politics in Japan after Fukushima," *Asian Currents*, 25 Jun. 2018, http://asaa.asn.au/anti-nuclear-movement-street-politics-japan-fukushima/. See also Alexander James Brown, *Anti-Nuclear Protest in Post-Fukushima Tokyo: Power Struggles* (Abingdon, Oxon: Routledge, 2018).
123. National Diet of Japan, Fukushima Nuclear Accident Independent Investigation Commission, *Report*, English language version (Tokyo: The National Diet, 2012).
124. National Diet of Japan, *Report*, English language version, 12. See also 15.
125. Kiyoshi Kurokawa, "The Fukushima Daiichi Nuclear Power Plant Accident Is Not Over," in National Diet of Japan, *Report*, English language version, 3–4.
126. National Diet of Japan, *Report*, English language version, chapter 5.
127. Kiyoshi Kurokawa, "Message from the Chairman," in *The Official Report of the Fukushima Nuclear Accident Independent Investigation Commission: Executive Summary*, ed. National Diet of Japan (Tokyo: National Diet of Japan, 2012), 9. Also quoted in Andy Horowitz, "Official Fukushima

Report Blames Japanese Culture, Not Nuclear Power," *The Atlantic*, 11 July 2012, https://www.theatlantic.com/international/archive/2012/07/official-fukushima-report-blames-japanese-culture-not-nuclear-power/259665/
128. Toshihiro Higuchi, "Japan's Culture: Culprit of the Nuclear Accident?" *Bulletin of the Atomic Scientists*, 4 Sept. 2012, https://thebulletin.org/2012/09/japans-culture-culprit-of-the-nuclear-accident/
129. Higuchi, "Japan's Culture."
130. Don Brown, "One Take on Japanese Cinema: Monsters vs. Gangsters vs. (Illegal) Aliens," *AJW (Asia & Japan Watch)*, 12 Aug. 2016, Dow Jones Factiva.
131. "Just Like the 1954 Original, New Godzilla Embodies Our Age of Anxiety," *The Mainichi*, 23 Aug. 2016, Dow Jones Factiva.
132. "Just Like the 1954 Original."
133. Tony Rayns, "Shin Godzilla," *Sight & Sound* (Oct. 2017): 74–75.
134. Peter Tasker, "Godzilla: A Constitutional Argument," *Nikkei Report*, 1 Dec. 2016, Dow Jones Factiva.
135. Anna Fifield, "The Real-World Roots of Godzilla," *Washington Post*, 23 Sept. 2016, EBSCOhost.
136. Tomoko Murakami and Venkatachalam Anbumozhi (eds), *An International Analysis of Public Acceptance of Nuclear Power*, ERIA (Economic Research Institute for ASEAN and East Asia) Research Project Report FY2017, no. 3 (Oct. 2018), 13–16, accessed 31 Jan. 2019, http://www.eria.org/publications/an-international-analysis-of-public-acceptance-of-nuclear-power/; Japan Atomic Energy Relations Organization, *2017 nendo genshiryoku ni kansuru seron chōsa* (*FY2017 Public Opinion Survey on Nuclear Power*) (Tokyo: JAERO, 2018), esp. 240–242, accessed 31 Jan. 2019, https://www.jaero.or.jp/data/01jigyou/tyousakenkyu29.html
137. Kei Yamada, "Overview of Fact-Finding Survey of the Japanese Nuclear Industry 2017 (through March 2018)," *Atoms in Japan*, 4 Feb. 2019, https://www.jaif.or.jp/en/overview-of-fact-finding-survey-of-the-japanese-nuclear-industry-2017-through-march-2018/
138. Government of Japan, METI (Ministry of Economy, Trade and Industry), *Strategic Energy Plan*, provisional translation, Jul. 2018, 3, accessed 31 Jan. 2019, http://www.enecho.meti.go.jp/en/category/others/basic_plan/5th/pdf/strategic_energy_plan.pdf
139. "METI's New Energy Agenda is Still Powered by Old Thinking," *Asahi Shimbun*, 18 May 2018, editorial, accessed 31 Jan. 2019 http://www.asahi.com/ajw/articles/AJ201805180028.html
140. "Inside Fukushima Daiichi," TEPCO, accessed 6 Feb. 2019, http://www.tepco.co.jp/en/insidefukushimadaiichi/index-e.html; Kyodo,

"TEPCO Offers English-Language Virtual Tour of Crippled Fukushima Nuclear Plant," *Japan Times*, 5 Nov. 2018, https://www.japantimes.co.jp/news/2018/11/05/national/tepco-offers-english-language-virtual-tour-crippled-fukushima-nuclear-plant/#.XFpei1wzaUk

141. Tim Kelly, with Kwiyeon Ha and Toru Hanai, "As Fukushima Residents Return, Some See Hope in Nuclear Tourism," World News, www.reuters.com, 21 Jun. 2018, https://www.reuters.com/article/us-japan-fukushima-nuclear-tourism/as-fukushima-residents-return-some-see-hope-in-nuclear-tourism-idUSKBN1JH081
142. Tomohiro Osaki, "Eight Years On, Abe Says 3/11 Recovery Nearing 'Final Stages,' Though Half of Public Unconvinced," *Japan Times*, 11 Mar. 2019, https://www.japantimes.co.jp/news/2019/03/11/national/eight-years-abe-says-3-11-reconstruction-nearing-final-stages-though-half-public-unconvinced/#.XJ11TfZuKUm
143. Tokyo Organizing Committee of the Olympic and Paralympic Games, *Tokyo 2020 Guidebook* (Tokyo: The Committee, 2017), 15.
144. Magdalena Osumi, "Torch Relay for 2020 Games to Start in Fukushima," *Japan Times*, 13 Jul. 2018, Dow Jones Factiva.
145. "'Games of Recovery': Tokyo Olympics Torch Relay to Start from Disaster-Hit Fukushima," Scroll.in, 12 July 2018, https://scroll.in/field/886229/games-of-recovery-tokyo-olympics-torch-relay-to-start-from-disaster-hit-fukushima
146. "About the Tokyo 2020 Olympic Torch Relay," Tokyo 2020, accessed 3 Jun. 2019, https://tokyo2020.org/en/special/torch/olympic/about/
147. "Tokyo 2020 Announces Torchbearer Application Opportunities and Unveils Uniform," The International Olympic Committee, accessed 1 Jun. 2019, https://www.olympic.org/news/tokyo-2020-announces-torchbearer-application-opportunities-and-unveils-uniform
148. "Tokyo 2020 Reveals Olympic Torch Design, Ambassadors and Relay Emblem," The International Olympic Committee, accessed 20 Mar. 2019, https://www.olympic.org/news/tokyo-2020-reveals-olympic-torch-design-ambassadors-and-relay-emblem
149. Reuters, "Olympics -'We Are Alive': Tokyo 2020 Ceremonies to Focus on Rebirth," *SBS News*, accessed 31 Jul. 2018, https://www.sbs.com.au/news/olympics-we-are-alive-tokyo-2020-ceremonies-to-focus-on-rebirth
150. Eric Johnston, "Osaka Kicks Off Preparations for 2025 Expo," *Japan Times*, 12 Dec. 2018, Dow Jones Factiva.
151. Eric Johnston, "Osaka Officials Visit Paris for Final 2025 Expo Pitch," *Japan Times*, 15 Jun. 2018, Dow Jones Factiva.

152. Eric Johnston, "Osaka Bids to Rekindle Magic of 1970 Expo But Taxpayers Doubt Lofty Plan's Claims," *Japan Times*, 24 Oct. 2016, Dow Jones Factiva.
153. Yoshimi Shunya, "A Drifting World Fair: Cultural Politics of Environment in the Local/Global Context of Contemporary Japan," in *Japan after Japan: Social and Cultural Life from the Recessionary 1990s to the Present*, eds. Tomiko Yoda and Harry Harootunian (Durham: Duke University Press, 2006), 395–414, esp. 396.
154. Richard J. Samuels, "Japan's 3.11 Master Narrative Still under Construction," *East Asia* Forum, 6 Mar. 2016, http://www.eastasiaforum.org/2016/03/06/japans-3-11-master-narrative-still-under-construction/
155. *FY 2012 Annual Report*, provisional English translation (Tokyo: Nuclear Regulation Authority, c. 2013), 2, accessed 6 Feb., 2019, http://www.nsr.go.jp/data/000067053.pdf
156. Ernest Renan, "What is a Nation?" (1882), trans. Martin Thom, in *Becoming National: A Reader*, eds. Geoff Eley and Ronald Grigor Suny (New York: Oxford University Press, 1996), 42–55, esp. 53.
157. Tessa Morris-Suzuki, "Touching the Grass: Science, Uncertainty and Everyday Life from Chernobyl to Fukushima," *Science, Technology & Society* 19, no. 3 (2014): 331–362, esp. 352. See also Tessa Morris-Suzuki, "Re-Animating a Radioactive Landscape: Informal Life Politics in the Wake of the Fukushima Nuclear Disaster," *Japan Forum* 27, no. 2 (2015): 167–188.
158. Yamada, "Overview of Fact-Finding Survey of the Japanese Nuclear Industry 2017 (through March 2018)."

Bibliography

"About the Tokyo 2020 Olympic Torch Relay," *Tokyo 2020*. Accessed 3 Jun. 2019. https://tokyo2020.org/en/special/torch/olympic/about/.

Acton, James M., and Mark Hibbs. *Why Fukushima Was Preventable*, The Carnegie Papers. Washington, DC: Carnegie Endowment for International Peace, 2012.

Aldhous, Peter, and Zena Iovino, "Japan's Record of Nuclear Cover-Ups." *New Scientist* 209, no. 2805 (26 March 2011): 11.

Aldrich, Daniel P. "Rethinking Civil Society-State Relations in Japan after the Fukushima Accident." *Polity* 45, no. 2 (April 2013): 249–264.

Alvarez, Robert. "No Bygones at Yongbyon," *Bulletin of the Atomic Scientists* 59, no. 4 (Jul.-Aug. 2003): 38–45.

Asahi Shimbun, 16 Jul. 2004.

Bermudez Jr. Joseph S. *North Korea's Development of a Nuclear Weapons Strategy*, North Korea's Nuclear Futures Series. Washington, D.C.: US-Korea Institute at SAIS, Johns Hopkins University, 2015.

Borggreen, Gunhild. "Ruins of the Future: Yanobe Kenji Revisits Expo '70." *Performance Paradigm* 2 (March 2006): 123–136.

Brown, Alexander. "The Anti-Nuclear Movement and Street Politics in Japan after Fukushima," *Asian Currents*, 25 Jun. 2018, http://asaa.asn.au/anti-nuclear-movement-street-politics-japan-fukushima/.

Brown, Alexander James. *Anti-Nuclear Protest in Post-Fukushima Tokyo: Power Struggles*. Abingdon, Oxon: Routledge, 2018.

Brown, Don. "One Take on Japanese Cinema: Monsters vs. Gangsters vs. (Illegal) Aliens," *AJW* (*Asia & Japan Watch*), 12 Aug. 2016, Dow Jones Factiva.

Burr, William, Barbara Elias and Robert Wampler, eds. "Nuclear Weapons on Okinawa Declassified December 2015, Photos Available since 1990." *Briefing Book* no. 541. Washington D.C.: National Security Archive, George Washington University, 19 Feb. 2016. https://nsarchive.gwu.edu/briefing-book/japan-nuclear-vault/2016-02-19/nuclear-weapons-okinawa-declassified-december-2015.

Clemens Jr. Walter C. "Korea's Quest for Nuclear Weapons: New Historical Evidence." *Journal of East Asian Studies* 10, no. 1 (Jan.-Apr. 2010): 127–154.

Dawson, Chester. "The Lighter Side of Plutonium," *WSJ Blogs*, 29 Mar. 2011, https://blogs.wsj.com/japanrealtime/2011/03/29/the-lighter-side-of-plutonium/.

"Denryokukan" ("Electric Power Pavilion"). *Expo '70 Commemorative Park*. Accessed 31 May 2019 https://www.expo70-park.jp/cause/expo/electric_power/.

Droubie, Paul. "Phoenix Arisen: Japan as Peaceful Internationalist at the 1964 Tokyo Summer Olympics." *International Journal of the History of Sport* 28, no., 16 (Nov. 2011): 2309–2322.

"Faults under Japan's Shika NPP Said to Be Active." *Nuclear Engineering International*, 4 Mar. 2016. Accessed 31 Jan. 2019 https://www.neimagazine.com/news/newsfaults-under-japans-shika-npp-said-to-be-active-4830171.

Feder, Toni. "Japan Arrests Six in Nuclear Accident that Killed Two." *Physics Today* 53, no. 12 (2000): 61-62.

"Follow-up Tokai-Mura and Leak at Fugen ATR," *Nuclear Monitor Issue*, no. 471 (25 Apr. 1997), https://www.wiseinternational.org/nuclear-monitor/471/follow-tokai-mura-and-leak-fugen-atr.

Friedman, Uri. "Hiroshima and the Politics of Apologizing." *The Atlantic* 26 May 2016, https://www.theatlantic.com/international/archive/2016/05/obama-hiroshima-apology-nuclear/483617/.

Fujimura, Joan. "Future Imaginaries: Genome Scientists as Sociocultural Entrepreneurs." In *Genetic Nature/Culture: Anthropology and Science beyond the Two-Culture Divide*, eds. Alan H. Goodman, Deborah Heath and M. Susan Lindee, 176–99. Berkeley: University of California Press, 2003.

FY 2012 Annual Report, provisional English translation (Tokyo: Nuclear Regulation Authority, c. 2013), 2, http://www.nsr.go.jp/data/000067053.pdf, accessed 6 Feb., 2019.

"Games of Recovery': Tokyo Olympics Torch Relay to Start from Disaster-Hit Fukushima," Scroll.in. Accessed 12 July 2018. https://scroll.in/field/886229/games-of-recovery-tokyo-olympics-torch-relay-to-start-from-disaster-hit-fukushima.

Gardner, William O. "The 1970 Osaka Expo and/as Science Fiction." *Review of Japanese Culture and Society* 28 (Dec. 2011): 26–43.

George Washington University, Digital National Security Archive.

Gibson, Alicia. "Out of Death, an Atomic Consecration to Life: *Astro Boy* and Hiroshima's Long Shadow." *Mechademia* 8, Tezuka's Manga Life (2013), 313–20.

"Gov. Azuma Pledges Full Support." *Japan Times*, 27 May 1959.

Government of Japan, METI (Ministry of Economy, Trade and Industry), *Strategic Energy Plan*, provisional translation, Jul. 2018, 3. Accessed 31 Jan. 2019 http://www.enecho.meti.go.jp/en/category/others/basic_plan/5th/pdf/strategic_energy_plan.pdf.

"Graphic Design of Kono Takashi," National Museum of Modern Art, Tokyo. Accessed 28 Apr. 2015. http://www.momat.go.jp/CG/KONO/.

Gross Levi, Barbara. "What happened at Tokaimura?" *Physics Today* 52, no. 12 (Dec. 1999): 52–54.

Hadfield, Peter. "Let's Have a Cup of Nice, Safe Plutonium." *New Scientist*, no. 1910 (29 Jan. 1994), https://www.newscientist.com/article/mg14119101-400-lets-have-a-cup-of-nice-safe-plutonium/.

Hamota, Hiroshi. "Olympics Flame Sent Message to World About New Postwar Japan." *Asahi Shimbun*, 31 Mar. 2003, Asahi Kikuzo II Visual database.

Haryū, Ichirō. "Expo '70 as the Ruins of Culture (1970)." Translated by Ignacio Adriasola. *Review of Japanese Culture and Society* 23, Expo '70 and Japanese Art: Dissonant Voices (Dec. 2011): 44–56.

Hashizume, Takehiro. "Kisha to genshiryoku" ("Atomic Energy As Seen by a Journalist"). *Nihon Genshiryoku Gakkaishi* 41, no. 2 (1999), 86–87.

Hecker, Siegfried S. "Report on North Korean Nuclear Program," Center for International Security and Cooperation, Stanford University, 15 Nov. 2006, https://cisac.fsi.stanford.edu/publications/report_on_north_korean_nuclear_program.

Hecker, Siegfried S., Sean C. Lee and Chaim Braun. "North Korea's Choice: Bombs over Electricity." *The Bridge: Linking Engineering and Society* (Summer 2010): 5–12.

Hecker, Siegfried S., Chaim Braun and Chris Lawrence, "North Korea's Stockpiles of Fissile Material." *Korea Observer* 47, no. 4 (Winter 2016): 721–749.

Higuchi, Toshihiro. "Japan's Culture: Culprit of the Nuclear Accident?" *Bulletin of the Atomic Scientists*, 4 Sept. 2012. https://thebulletin.org/2012/09/japans-culture-culprit-of-the-nuclear-accident/.

Hoey, Fintan. *Satō, America and the Cold War: U.S.-Japanese Relations 1964–72.* Houndmills, Basingstoke: Palgrave Macmillan, 2015.

Hook, Glenn D. "The Nuclearization of Language: Nuclear Allergy as Political Metaphor," *Journal of Peace Research* 21, no. 3 (1984): 259–75.

Horowitz, Andy. "Official Fukushima Report Blames Japanese Culture, Not Nuclear Power." *The Atlantic*, 11 Jul. 2012. https://www.theatlantic.com/international/archive/2012/07/official-fukushima-report-blames-japanese-culture-not-nuclear-power/259665/.

Igarashi, Yoshikuni. "The Bomb, Hirohito, and History: The Foundational Narrative of United States-Japan Postwar Relations." *positions* 6, no. 2 (Fall 1998): 261–302.

"Inside Fukushima Daiichi." *TEPCO*. Accessed 6 Feb. 2019 http://www.tepco.co.jp/en/insidefukushimadaiichi/index-e.html.

Jameson, Samuel. "Japan Top Expo Attraction," *Chicago Tribune*, 22 Mar. 1970. http://archives.chicagotribune.com/1970/03/22/page/46/article/japan-top-expo-attraction.

Japan and the Japanese: The Japan Pavilion, English language version. Tokyo: Ministry of International Trade and Industry and the Japan External Trade Organization, 1970.

Japan Atomic Energy Relations Organization *2017 nendo genshiryoku ni kansuru seron chōsa (FY2017 Public Opinion Survey on Nuclear Power)*. Tokyo: JAERO 2018. Accessed 31 Jan. 2019. https://www.jaero.or.jp/data/01jigyou/tyousakenkyu29.html.

Japan Times, 13 Jul. 2018.

Japan Travel Bureau. *Your Technical Tour in Japan.* Tokyo: Nihon Keizai Shimbun, 1965.

Japanese Government, *Nihon to Nihonjin: Nihonkan (Japan and the Japanese: The Japan Pavilion)*. Tokyo: Ministry of International Trade and Industry and the Japan External Trade Organization 1970.

Johnson, Jesse. "In First, U.S. Admits Nuclear Weapons Were Stored in Okinawa during Cold War," *Japan Times*, 20 Feb. 2016, https://www.japantimes.co.jp/news/2016/02/20/national/history/first-u-s-admits-nuclear-weapons-stored-okinawa-cold-war/#.XOYjg-gzaUk

Johnston, Eric. "Osaka Bids to Rekindle Magic of 1970 Expo But Taxpayers Doubt Lofty Plan's Claims." *Japan Times*, 24 Oct. 2016, Dow Jones Factiva.

Johnston, Eric. "Osaka Kicks Off Preparations for 2025 Expo." *Japan Times*, 12 Dec. 2018a, Dow Jones Factiva.

Johnston, Eric. "Osaka Officials Visit Paris for Final 2025 Expo Pitch." *Japan Times*, 15 Jun. 2018b, Dow Jones Factiva.

Jōmaru, Yōichi, *Genpatsu to media: Shimbun jyānarizumu nidome no haiboku (Nuclear Power Generation and the Media: The Second Failing of Newspaper Journalism*. Tokyo: Asahi Shimbun Publications 2012.

"Just Like the 1954 Original, New Godzilla Embodies Our Age of Anxiety," *The Mainichi*, 23 Aug. 2016, Dow Jones Factiva.

"Kangeki no 'Kamoshika kun'" ("Deeply Moved Young Man Nicknamed 'Japanese Antelope'"), *Asahi Shimbun*, 11 Aug. 1964, evening edition.

Kelly, Tim, with Kwiyeon Ha and Toru Hanai. "As Fukushima Residents Return, Some See Hope in Nuclear Tourism," World News, www.reuters.com. 21 Jun. 2018. https://www.reuters.com/article/us-japan-fukushima-nuclear-tourism/as-fukushima-residents-return-some-see-hope-in-nuclear-tourism-idUSKBN1JH081

Kinefuchi, Etsuko. "Nuclear Power for Good: Articulations in Japan's Nuclear Power Hegemony." *Communication, Culture & Critique* 8 (2015): 448–465.

Kondō, Motohiro, "Japanese Creativity: Robots and *Anime*," *Japan Echo* 30, no. 4 (Aug. 2003): 6–8.

Kumagai, Jean. "In the Wake of Tokaimura, Japan Rethinks Its Nuclear Future." *Physics Today* 52, no. 12 (Dec. 1999): 51–52.

Kurokawa, Kiyoshi."Message from the Chairman." In *The Official Report of the Fukushima Nuclear Accident Independent Investigation Commission: Executive Summary*. Edited by National Diet of Japan. Tokyo: National Diet of Japan, 2012.

Kuwajima Kisha (Reporter Kuwajima). "Taiyaku o hatashita Sakai-kun" ("Young Sakai Who Played a Big Role"). *Asahi Shimbun*, 11 Oct. 1964, 14.

Kyodo, "TEPCO Offers English-Language Virtual Tour of Crippled Fukushima Nuclear Plant," *Japan Times*, 5 Nov. 2018, https://www.japantimes.co.jp/news/2018/11/05/national/tepco-offers-english-language-virtual-tour-crippled-fukushima-nuclear-plant/#.XFpei1wzaUk.

"Learning from the Lessons of 3/11, Seven Years On." *Japan Times*, 10 Mar. 2018, Dow Jones Factiva.

Lindee, Susan. "Survivors and Scientists: Hiroshima, Fukushima, and the Radiation Effects Research Foundation 1975–2014," *Social Studies of Science* 46, no. 2 (2016): 184–209, esp. 199.

Lindley, P.A., and K.J. Mitchell. "Engineering Design." *Nuclear Power* (March 1960): 104–107.

Lowry, David. "What Theresa May Forgot: North Korea Used British Technology to Build Its Nuclear Bombs." *Ecologist: The Journal for the Post-Industrial Age*, 26 Jul. 2016, https://theecologist.org/2016/jul/26/what-theresa-may-forgot-north-korea-used-british-technology-build-its-nuclear-bombs.

Mainichi, 3 Mar. 2016.

Manabe, Noriko. "Monju-kun: Children's Culture as Protest." In *Child's Play: Multi-Sensory Histories of Children and Childhood in Japan*. Edited by Sabine Frühstück and Anne Walthall, 264–285. Berkeley: University of California Press, 2017.

Matsuura, Masayoshi, Kohei Hisamochi, Shinichiro Sato and Kumiaki Moriya, "Lessons Learned from Fukushima Daiichi Nuclear Power Station Accident and Consequent Safety Improvements." *Hitachi Review* 62, no. 1 (2013): 75–80.

"METI's New Energy Agenda is Still Powered by Old Thinking," *Asahi Shimbun*, 18 May 2018, editorial. Accessed 31 Jan. 2019 http://www.asahi.com/ajw/articles/AJ201805180028.html.

"Milestones: Kurobe River No. 4 Hydropower Plant 1956–63." *Engineering and Technology History Wiki*, c. 2015. Accessed 18 Jan. 2019 https://ethw.org/Milestones:Kurobe_River_No._4_Hydropower_Plant,_1956-63.

Miyake, Toshio. "Popularising the Nuclear: Mangaesque Convergence in Postwar Japan." In *Rethinking Nature in Contemporary Japan: Science, Economics, Politics*. Edited by Marcella Mariotti, Toshio Miyake and Andrea Revelant, 71–93. Venice: Edizioni Ca' Foscari, 2014.

Miyazaki, Yoshikazu. "The Japanese-Type Structure of Big Business." In *Industry and Business in Japan*. Edited by Kazuo Sato, 285–343. White Plains, New York: M.E. Sharpe, 1980.

Moritaki, Ichirō. *Kaku to jinrui wa kyōzon dekinai: Kaku zettai hitei e no ayumi (Nuclear Weapons and Human Being Cannot Co-Exist: Steps Towards the Absolute Negation of Nuclear Weapons)*. Tokyo: Nanatsumori Shokan, 2015.

Morris-Suzuki, Tessa. "Re-Animating a Radioactive Landscape: Informal Life Politics in the Wake of the Fukushima Nuclear Disaster." *Japan Forum* 27, no. 2 (2015): 167–188.

Morris-Suzuki, Tessa. "Touching the Grass: Science, Uncertainty and Everyday Life from Chernobyl to Fukushima." *Science, Technology & Society* 19, no. 3 (2014): 331–362.

Murakami, Tomoko, and Venkatachalam Anbumozhi (eds), *An International Analysis of Public Acceptance of Nuclear Power*, ERIA (Economic Research Institute for ASEAN and East Asia) Research Project Report FY2017, no. 3 (Oct. 2018). Accessed 31 January 2019. http://www.eria.org/publications/an-international-analysis-of-public-acceptance-of-nuclear-power/.

National Diet of Japan, Fukushima Nuclear Accident Independent Investigation Commission. *Hōkokusho* (*Report*). Tokyo: The National Diet, 2012a.

National Diet of Japan, Fukushima Nuclear Accident Independent Investigation Commission. *Report*, English language version. Tokyo: The National Diet, 2012b.

New York Times, 23 Aug. 1964, 14 Nov. 1964, 1 Oct. 1999, 2–3 Dec. 1999, 16 Sept. 2002, 17 Oct. 2006, 25 Jun. 2011, 19 Mar. 2018.

New York Times Magazine, 29 Aug. 1965.

Norris, Robert S. "U.S. Weapons Secrets Revealed." *Bulletin of the Atomic Scientists* 49, no. 2 (Mar. 1993): 48.

"Nuclear Power in Fukui," Fukui Prefectural Environmental Radiation Research and Monitoring Centre. Accessed 7 Feb. 2019. http://www.houshasen.tsuruga.fukui.jp/en/pages/radiation/plant/plant2.html.

Official Report of the Japan World Exposition, Osaka 1970, Vol. 1 (Osaka: Commemorative Association for the Japan World Exposition (1970), 1972).

Oka, Takashi. "The New Japan." *The Rotarian* 97, no. 5 (Nov. 1960): 15–23.

"Ōsaka banpaku Nihon kan ni tenji no tapesutorii 36 nenburi ni kōkai" ("Osaka Expo Japan Pavilion's Tapestries on Public Display Again after 36 Years"). 8 Sept. 2006. http://www.asahi.com/culture/news_culture/OSK200609080052.html.

Osaki, Tomohiro. "Eight Years On, Abe Says 3/11 Recovery Nearing 'Final Stages,' Though Half of Public Unconvinced." *Japan Times*, 11 Mar. 2019. https://www.japantimes.co.jp/news/2019/03/11/national/eight-years-abe-says-3-11-reconstruction-nearing-final-stages-though-half-public-unconvinced/#.XJ11TfZuKUm.

"Political Meltdown," *The Economist* (17 Apr. 1997), https://www.economist.com/asia/1997/04/17/political-meltdown.

Pollock, Joshua. "Why Does North Korea Have a Gas-Graphite Reactor?" *Arms Control Wonk*, 16 Oct. 2009. https://www.armscontrolwonk.com/archive/502504/why-does-north-korea-have-a-gas-graphite-reactor/.

Power Reactor and Nuclear Fuel Development Corporation. *Purutoniumu monogatari: Tayoreru nakama Puruto-kun* (*Plutonium Story: Plutonium Boy, Our Reliable Friend*) (1993), uploaded as "Japanese Nuclear Propaganda Cartoon." Accessed 10 Jan. 2020 https://www.youtube.com/watch?v=Iw1LYthC4PQ&t=8s.

Rayns, Tony. "Shin Godzilla." *Sight & Sound* (Oct. 2017): 74–75.

Renan, Ernest. "What is a Nation?" (1882). Translated by Martin Thom. In *Becoming National: A Reader*. Edited by Geoff Eley and Ronald Grigor Suny, 42–55. New York: Oxford University Press, 1996.

Reuters. "Olympics -'We Are Alive': Tokyo 2020 Ceremonies to Focus on Rebirth," *SBS News*, 31 Jul. 2018. https://www.sbs.com.au/news/olympics-we-are-alive-tokyo-2020-ceremonies-to-focus-on-rebirth.

Ryan, Michael E. "The Tokaimura Nuclear Accident: A Tragedy of Human Errors." *Journal of College Science Teaching* 31, no. 1 (Sept. 2001): 42–48.

Sabanovic, Selma. "Inventing Japan's 'Robotics Culture': The Repeated Assembly of Science, Technology, and Culture in Social Robotics." *Social Studies of Science* 44, no. 3 (2014): 342–67.

Sakai, Yoshinori. "Wakai sedai no eiyo" ("Young Generation's Honour"). *Asahi Shimbun*, 19 Aug. 1964.

Samuels, Richard J. "Japan's 3.11 Master Narrative Still under Construction." *East Asia* Forum, 6 Mar. 2016. http://www.eastasiaforum.org/2016/03/06/japans-3-11-master-narrative-still-under-construction/.

Schodt, Frederik. *The Astro Boy Essays: Osamu Tezuka, Mighty Atom, and the Manga/Anime Revolution*. New York: Stone Bridge Press, 2007.

"Seika saishū rannā" ("Final Torchbearer"), *Shūkan shinchō* (*Weekly Shinchō*), 14–21 August 2008, 62.

Shikama, Kōichi. "'Banpaku ni genshi no akari o', hairo de omoidashita Mihama shuzai" ("'Atomic Light for the Expo' and Coverage at the Decommissioned Nuclear Reactor at Mihama." *Sankei Shimbun*, 19 Mar. 2015. http://www.sankei.com/west/news/150319/wst1503190023-n1.html;

"Shinku no honō" ("Hot Crimson Flame"), *Asahi Shimbun*, 10 October 1964, evening edition.

Smith, Aileen Mioko. "Nuclear Mayhem at Monju." *Earth Island Journal* 11, no. 2 (Spring 1996): 22.

Smith, Sandy. "JCO Employees Plead Guilty to Negligence in Deaths at Japanese Nuclear Facility." *Safety Online*, 24 Apr. 2001. https://www.safetyonline.com/doc/jco-employees-plead-guilty-to-negligence-in-d-0001.

Sports Illustrated, Oct. 19, 1964, cover.

Steinberg, Marc. "Anytime, Anywhere: *Tetsuwan Atomu* Stickers and the Emergence of Character Merchandizing." *Theory, Culture & Society* 26 (2–3): 113–138.

Storry, Richard. "Defence and Economic Links with America." *Financial Times*, 29 Nov. 1967.

Sumihara, Noriya. "Flamboyant Representation of Nuclear Power Station Visitor Centers in Japan: Revealing or Concealing, Or Concealing by Revealing?" *Agora: Journal of International Center for Regional Studies* 1 (2003): 11–29.

Swinbanks, David. "Sodium Leak Blots Japan's Nuclear Prospects." *Nature* 379, no. 6562 (18 Jan. 1996): 196.

Tachibana, Takashi. "'Robotto ōkoku' Nihon no mōten" ("The Blind Spot of Robot Kingdom Japan"). *Bungei shunju* 81, no. 6 (May 2003): 144–159.

Takahata, Seiichi. *Industrial Japan and Industrious Japanese*. Osaka: Nissho Co., 1968.

"Takashi Kono," *Idea No. 60*, Tenth Commemorative Special Issue (Aug. 1963). Accessed 28 Apr. 2015. http://www.idea-mag.com/en/publication/060.php.

"Takashi Kono's Profile," Gallery 5610. Accessed 28 Apr. 2015, http://www.deska.jp/profile.

Takubo, Masa. "Closing Japan's Monju Fast Breeder Reactor: The Possible Implications." *Bulletin of the Atomic Scientists* 73, no. 3 (2017): 182–187.

Tanaka, Yuki. "War and Peace in the Art of Tezuka Osamu: The Humanism of His Epic Manga." *Asia-Pacific Journal* 8, issue 38, no. 1 (20 Sept. 2010): 1–15.

Tasker, Peter. "Godzilla: A Constitutional Argument." *Nikkei Report*, 1 Dec. 2016, Dow Jones Factiva.

"Tokyo Gets '64 Olympic Games." *Japan Times*, 27 May 1959.

"Tokyo 2020 Announces Torchbearer Application Opportunities and Unveils Uniform, The International Olympic Committee", 1 Jun. 2019. https://www.olympic.org/news/tokyo-2020-announces-torchbearer-application-opportunities-and-unveils-uniform.

"Tokyo 2020 Reveals Olympic Torch Design, Ambassadors and Relay Emblem," The International Olympic Committee. Accessed 20 Mar. 2019. https://www.olympic.org/news/tokyo-2020-reveals-olympic-torch-design-ambassadors-and-relay-emblem.

Tokyo Organizing Committee of the Olympic and Paralympic Games. *Tokyo 2020 Guidebook*. Tokyo: The Committee, 2017.

Umehara, Takeshi. "Genshiryoku wa Nihon no kami ni niteiru: Kōdo na ningen no chie de chinkon o" ("Nuclear Power Is Like a God to Japan: Advanced Knowledge Brings About the Repose of Souls"), *Genshiryoku bunka (Nuclear Culture)* 17, no. 8 (August 1986): 3–8.

Washington Post, 23 Sept. 2016.

Watanabe, Ichirō, ed. *Tōkai Hatsudensho no kensetsu: Genshiryoku hatsuden paionia no kiroku (The Construction of the Tōkai Power Station: A Record of a Nuclear Power Pioneer)*. Tokyo: Japan Atomic Power Company, 1971.

"World: Toward the Japanese Century," *Time* (2 Mar. 1970), http://content.time.com/time/magazine/article/0,9171,904215,00.html.

Yamada, Kei. "Overview of Fact-Finding Survey of the Japanese Nuclear Industry 2017 (through March 2018)." *Atoms in Japan*, 4 Feb. 2019. https://www.jaif.or.jp/en/overview-of-fact-finding-survey-of-the-japanese-nuclear-industry-2017-through-march-2018/.

"Yongbyon 5MWe Reactor," Nuclear Threat Initiative, 19 July 2018, https://www.nti.org/learn/facilities/766/.

Yoshida, Reiji, and Takahiro Fukada, "Fukushima Plant Site Originally was a Hill Safe from Tsunami," *Japan Times*, 13 Jul. 2011, Dow Jones Factiva.

Yoshimi, Shunya. "A Drifting World Fair: Cultural Politics of Environment in the Local/Global Context of Contemporary Japan." In *Japan after Japan: Social and Cultural Life from the Recessionary 1990s to the Present*. Edited by Tomiko Yoda and Harry Harootunian, 395–414. Durham: Duke University Press, 2006.

Yoshimi, Shunya. *Yume no genshiryoku (Atoms for Dreams)*. Tokyo: Chikuma Shobō, 2012.

Yoshioka, Hitoshi. "The Rise and Fall of Fast Breeder Development in Japan," *Science Studies* 9, no. 2 (1996): 14–26.

CHAPTER 10

Conclusion

Looking Back

In this book, we have seen how the Japanese have visualized nuclear power by means of their imagination but also the ideas and images provided to them by governments, scientists and politicians, media outlets and power companies, as well as photographers, film-makers, artists and writers. This book made clear that they often did not act alone but in conjunction with others. The media was particularly important in Japan in facilitating the promotion of nuclear power. People obtained information about nuclear power from newspapers, magazines, *kamishibai* performed in the streets, and exhibitions held in consumer-friendly department stores. These media-intensive efforts served to naturalize nuclear power, making it a part of everyday Japanese life.

From the time of the decision to use the atomic bomb on Hiroshima in 1945, visual thinking has been important in the process. But the promotion of nuclear power in Japan has been more than just pictures. It was very much a multi-sensory story where the visual and experiential worked together with narratives and discourses about the role of science and technology in Japan's future. Chapters in this book went into considerable detail about how visual images, artefacts and models of reactors worked in tandem to impress and educate members of the public who saw and experienced what the future would look like.

M. Low, *Visualizing Nuclear Power in Japan*, Palgrave Studies in the History of Science and Technology,
https://doi.org/10.1007/978-3-030-47198-9_10

Visualizing nuclear power involved sensory feedback from the whole body and the story of *Alice in Atom-Land* that featured in some exhibitions highlights this idea of literally being transported to a different place and time. Like Alice, the Japanese people could gain a sense of what it was like to visit a reactor in the atomic age. The public were moved by what they saw. In a way, the many exhibitions, fairs, expos and international events that the Japanese people participated in conveyed visual stories that sought to shape who the Japanese people were or hoped to be.

Some individuals were singled out in this book for special attention. Frances Baker (later known as Frances Blakemore) and Akamatsu Toshiko (also known as Maruki Toshi) both had a talent for visual expression which they used to their advantage during World War II, during the Allied Occupation and in subsequent years. Baker was highly active not only in wartime propaganda but also in CIE exhibits during the Occupation and after, USIS-sponsored events.

The media mogul Shōriki Matsutarō saw the promotion of nuclear power as a great opportunity for Japan and, with the help of politicians such as Nakasone Yasuhiro, sought to realize Japan's nuclear dreams. This involved skilful use of the media. Scientists such as Yukawa Hideki and Clark D. Goodman were other key figures. Their expertise in nuclear physics was also an integral part of the story with even a film being made about Yukawa's visit to the USA and how it enhanced US-Japan relations.

At trade fairs and conferences, Americans vied with the British to sell nuclear technology to the Japanese. Models and diagrams of reactors gave scientists, industrialists and politicians a taste of what was to come. We discussed films showing nuclear power plants under construction. They served to further aestheticize Japan's commitment to nuclear power and conveyed a sense that Japanese ingenuity (with foreign assistance) had made the atomic age visible by taming the atom for the benefit of the nation.

Eventually, when reactors were built, members of the public including students could experience the real thing themselves. But even then, orientation was required at museums and visitor centres in order to make what visitors saw meaningful. In a way, reactors became sensory spaces where visitors (including students) and technical staff performed their roles as part of the national effort to make Japan more self-sufficient in energy. Nuclear reactors were presented as safe and integral to Japan's post-war economic growth.

The effectiveness of the promotion of nuclear power was linked to not only showing and demonstrating but also in hiding and learning not to see.[1] We saw manifestations of this in atomic bomb censorship during the Occupation, in *The Family of Man* exhibition where Yamahata Yōsuke's photomural was hidden from view, and in nuclear accidents where the severity of the situation was often downplayed. The desire to embrace nuclear power seemed to make people disregard the potential dangers.

Yet at the same time, there were countervailing forces such as the Hiroshima panels and nuclear-themed films that challenged the authority of those promoting Atoms for Peace. Artists and activists reminded the Japanese people what had occurred at Hiroshima, Nagasaki, Bikini Atoll and elsewhere. The public responses were visceral and reflected growing nuclear fear, especially in the wake of the Lucky Dragon Incident.

Visual Activism

As we saw with Akamatsu and her husband Maruki Iri, active engagement was required to effect change and bring greater awareness of what had happened in August 1945. Today, we could describe the activities of the Marukis as a form of visual activism. We should also note that the exhibition of the Hiroshima panels involved performative elements such as talks by students who sought to explain and build on issues raised by the mural-sized paintings.[2] Collectively, they were able to counter the way in which the government and the press often depicted atomic energy as a power for good.

Even in post-Occupation Japan before the spread of television, we can point to the role of the magazine *Asahi Gurafu* (*Asahi Graph*) in providing televisual reporting of events such as the Lucky Dragon Incident. Freed from the constraints of press censorship during the Occupation, we could also describe its coverage of events as a type of visual activism. Even then, there was an awareness that the visual opened up avenues for creating social and political change.

Writing for *The New Yorker*, Carolyn Kormann interviewed the Tohoku University-based expert on energy and climate policy Asuka Jusen who suggested that Japan was not generally a protest culture and bringing about change in Japan was difficult.[3] Some young Japanese activists had recently brainstormed about how they could bring more attention to climate change. Asuka was exasperated by a young woman's suggestion that, as a mark of protest, they could walk more slowly at the famous Shibuya

Crossing where five roads meet. Adjacent to Shibuya Station, this iconic intersection is flanked by three huge television screens that are mounted on buildings. The screens show images constantly and waves of people converge on to the crossing when the green signal for pedestrians lights up and flashes. Depicted in well-known foreign films such as *Lost in Translation* (2003) and *The Fast and the Furious: Tokyo Drift* (2006), going slow here at one of the most hectic crossings in Tokyo would have sent a message that the pace of life in Japan is too hectic, that the Japanese consume too much energy, and that some things taken for granted in everyday life need to change.

Cute Direct Action

Kormann herself relates how she met one woman at an Earth Day festival in Tokyo who introduced her to a green bear mascot who duly handed her his business card. The mascot, known as Zeronomikuma ("kuma" means bear in Japanese) was promoting "Zeronomics" which advocated zero reliance on fossil fuels and nuclear power, in contrast with Prime Minister Abe Shinzō's policies known as "Abenomics" which sought to revive the Japanese economy with a number of policies which were reliant on "cheap" energy provided by nuclear power. The green furry bear had a black zero on its belly, reminding all who met him that he was advocating an economy that did not rely on atomic energy.[4]

While Kormann and Asuka may have considered such initiatives as lacking in aggression and rather tame, they are examples nevertheless of how the Japanese people are seeking to use popular culture to challenge official narratives about nuclear power and its impact on the environment. Walking more slowly at a famous pedestrian crossing was a creative way to convey through the body that Japan needed a change in priorities. And as we have seen with Astro Boy who was powered by a nuclear reactor and Godzilla who became a mutant monster thanks to nuclear testing, cute characters can be very persuasive. What this reminds us is that the very same methods used by pro-nuclear bodies in characters such as Plutonium Boy can also be used to throw light on alternative views of nuclear power.

Manabe Noriko has written about "Monju-kun," a cartoon character in the form of a cute and sickly boy who was modelled on the ailing Monju fast breeder reactor in Fukui prefecture that has been shut down and will be decommissioned over the next thirty years. The mascot-like character emerged in the wake of the Fukushima nuclear disaster. He made the scary problem of nuclear reactors visible in a more palatable form. While seemingly only of appeal to children and perhaps their mothers, even older Japanese seem to enjoy learning from such fictional characters.[5] The use of cute characters enables the Japanese people to contemplate what they otherwise might not want to see. We may laugh at how idealistic some of this might appear to be, but such minor actions could build into calls for greater social and political change that might challenge the status quo.

Alexander Brown has written about the cultural practices of the anti-nuclear movement that include the skilful use of urban space, media production, music, visual art, street theatre and lively intellection debate. He mentions how activist groups such as Sayonara Genpatsu (Goodbye Nuclear Power Plants) held rallies and demonstrations in mid-2012 where Monju-kun and others appeared. In this way, members of the public were engaged in a lively way.[6]

One aspect of Tokyo protests has been what he calls anti-nuclear "sound demonstrations" involving bands, DJs and rappers performing on the back of trucks. Through their music, they are able to express their emotions and those attending can join in and move their bodies according to their own rhythm. This creates a festive atmosphere of music, dance and irreverent protest which flows through to other live music venues, cafes, bookshops and bars, forming what can be called a culture of rebellion.[7]

Japan arguably does have a protest culture but its forms differ from what has traditionally been regarded as political protest. The anti-nuclear movement has tended towards non-violent protests that often incorporate artistic expression whether it be in music, dance or visual art. Gonoi Ikuo calls this aesthetic approach "*kawaii* (cute) direct action."[8] He considers this art-centred approach more feminine than the masculine, more violent approach to resistance that can be observed in other parts of the world. He cites the endearing, cute image of Monju-kun as an example of this approach.

This chapter marks the end of this book but not necessarily the end of the story of nuclear power in Japan. Whereas nuclear power accounted for approximately one-third of Japan's total domestic electrical power generation before the Fukushima nuclear disaster, it dropped to zero in 2012 when all Japan's nuclear power plants were effectively offline or shut down. With the gradual restart of some reactors, nuclear power would make up 2.8 per cent of total electricity generation in 2017, increasing to 4.7 per cent in 2018.[9] Whether it will return to previous highs is a matter for the Japanese people and their representatives to decide and to act on.

What is clear is that aesthetic forms of protest have been and will continue to be an integral part in the ongoing story of nuclear power in Japan.

Notes

1. Nicholas Mirzoeff, *How to See the World* (Harmondsworth: Pelican Books, 2015), 14.
2. Mirzoeff, *How to See the World*, esp. 289–298.
3. Carolyn Kormann, "Is Nuclear Power Worth the Risk?" *New Yorker*, 22 Dec. 2019, https://www.newyorker.com/news/dispatch/is-nuclear-power-worth-the-risk
4. Anna Wiemann, *Networks and Mobilization Processes: The Case of the Japanese Anti-Nuclear Movement after Fukushima* (Tokyo: German Institute for Japanese Studies, 2018), 147.
5. Noriko Manabe, "Monju-kun: Children's Culture as Protest," in *Child's Play: Multi-Sensory Histories of Children and Childhood in Japan*, eds. Sabine Frühstück and Anne Walthall (Berkeley: University of California Press, 2017), 264–285.
6. Alexander Brown, *Anti-Nuclear Protest in Post-Fukushima Tokyo: Power Struggles* (Abingdon, Oxon: Routledge, 2018), esp. 148–49.
7. Alexander James Brown, "Above and Below the Streets: A Musical Geography of Anti-Nuclear Protest in Tokyo," *Emotion, Space and Society* 20 (2016): 82–89.
8. Ikuo Gonoi, "'The Cloudization' of Social Movements: The Esthetic Approach to Protest through the Example of '*Kawaii* Direct Action'," *Japan Political Science Review* 2 (2014): 1–17.
9. Institute for Sustainable Energy Policies, Tokyo, "Share of Renewable Energy Power in Japan, 2018 (Preliminary Report)," 5 Apr. 2019, accessed 13 Feb. 2020, https://www.isep.or.jp/en/717/

Bibliography

Brown, Alexander. *Anti-Nuclear Protest in Post-Fukushima Tokyo: Power Struggles.* Abingdon, Oxon: Routledge, 2018.

Brown, Alexander James. "Above and Below the Streets: A Musical Geography of Anti-Nuclear Protest in Tokyo." *Emotion, Space and Society* 20 (2016): 82–89.

Gonoi, Ikuo. "'The Cloudization' of Social Movements: The Esthetic Approach to Protest through the Example of '*Kawaii* Direct Action'." *Japan Political Science Review* 2 (2014): 1–17.

Institute for Sustainable Energy Policies, Tokyo. "Share of Renewable Energy Power in Japan, 2018 (Preliminary Report)." 5 Apr. 2019. Accessed 13 Feb. 2020. https://www.isep.or.jp/en/717/.

Kormann, Carolyn. "Is Nuclear Power Worth the Risk?" *New Yorker*, 22 Dec. 2019. https://www.newyorker.com/news/dispatch/is-nuclear-power-worth-the-risk.

Manabe, Noriko. "Monju-kun: Children's Culture as Protest." In *Child's Play: Multi-Sensory Histories of Children and Childhood in Japan*. Edited by Sabine Frühstück and Anne Walthall, 264-285. Berkeley: University of California Press, 2017.

Mirzoeff, Nicholas. *How to See the World*. London: Pelican Books, 2015.

Wiemann, Anna. *Networks and Mobilization Processes: The Case of the Japanese Anti-Nuclear Movement after Fukushima*. Tokyo: German Institute for Japanese Studies, 2018.

Index

M. Low, *Visualizing Nuclear Power in Japan*, Palgrave Studies in the History of Science and Technology,
https://doi.org/10.1007/978-3-030-47198-9

U

V

W

X

Y

Z